PLANT SCIENCE RESEARCH AND PRACTICES

AF735587

GINKGO BILOBA

BIOLOGY, USES AND HEALTH BENEFITS

PLANT SCIENCE RESEARCH AND PRACTICES

Additional books in this series can be found on Nova's website under the Series tab.

Additional e-books in this series can be found on Nova's website under the e-book tab.

PLANT SCIENCE RESEARCH AND PRACTICES

GINKGO BILOBA

BIOLOGY, USES AND HEALTH BENEFITS

EMMETT FISHER

EDITOR

NOVINKA

New York

NOTICE TO THE READER

Additional color graphics may be available in the e-book version of this book.

Library of Congress Cataloging-in-Publication Data
Names: Fisher, Emmett, editor.
Title: Ginkgo biloba : Biology, Uses and Health Benefits/ editor: Emmett Fisher.
Other titles: Plant science research and practices.
Description: Hauppauge, New York : Nova Science Publishers, [2016] | Series: Plant science research and practices | Includes bibliographical references and index.
Identifiers: LCCN 2015048135 (print) | LCCN 2016000019 (ebook) | ISBN 9781634844604 (softcover) | ISBN 9781634844857 ()
Subjects: LCSH: Ginkgo. | Ginkgo--Therapeutic use.
Classification: LCC QK494.5.G48 G56 2016 (print) | LCC QK494.5.G48 (ebook) | DDC 615.3/257--dc23
LC record available at http://lccn.loc.gov/2015048135

Published by Nova Science Publishers, Inc. † New York

CONTENTS

PREFACE

The use of medicinal plants or natural products increased in the last decades all over the world. One of the most popular herbal plants is *Ginkgo biloba* L. because of its widespread healing effects. Ginkgo has been used by the traditional medicine for thousands of years. It has been a sacred tree, a symbol of yin and yang, of longevity and vitality. The Chinese had begun their medicinal use of the tree with the seeds, but they soon added the leaves and even the roots to their list of valuable medicinal materials. *Ginkgo biloba* has been thoroughly investigated for its constituents, and a whole array of compounds has been described. Chapter One in this book presents the basic biological description, phylogeographic history, and the ecological characteristics of *Ginkgo biloba*. Chapter Two studies the elemental composition of *Ginkgo biloba* L. leaves. Chapter Three discusses different ginkgo preparations. Chapter Four presents the results of a screening of different crude extracts, infusions and standardized extract from the *Ginkgo biloba* (Ginkgoaceae) leaves for total phenolic content, concentration of flavonoids and in vitro antioxidant activity.

Chapter 1 - In this chapter, the basic biological description, phylogeographic history, as well as ecological characteristics of *Ginkgo biloba* are presented. One of the four divisions within Gimnospermae is the division Ginkgophyta. Ginkgophyta includes only the class Ginkgopsida, with the order Ginkgoales, which, in addition to four extinct families, encompasses a recent family Ginkgoaceae. The Ginkgoaceae family includes only one recent genus – *Ginkgo,* with the species *Ginkgo biloba*. The diversification of this genus began during the Jurassic and the early Cretaceous period. The first extinction of this species started in the polar regions at the end of the Miocene. Some parts of China have acquired the status of refugia for a "living fossil species" *Ginkgo biloba.* More scientific and common names exist for this species; among others, Carl Linnaeus constructed its Latin name as *Ginkgo biloba.* The height of *Ginkgo biloba* tree ranges between 20 and 30 m, with a diameter from 50 cm to 1 m or more. Shoots are differentiated into two distinct morphological types: long shoots with widely spaced leaves that subtend axillary buds; and short, or spur, shoots with clustered leaves that lack both internodes and axillary buds. The leaves have open dichotomous venation and are wedge-shaped with a semicircular undulate margin and a middle notch that divides the leaf surface into two halves. *Ginkgo biloba* is a dioecious plant. The male reproductive structures – microstrobiles are catkin-like, pollen-bearing cones, while the female reproductive structures – macrostrobiles, consist of a stalk-like peduncle and a pair of ovules. In the ovule begins the four-month long gametophyte development process, which results in the formation of two multiflagellated motile spermatozoids. After the egg cell formation, one of the motile spermatozoids carries out the process of fertilization, after which it performs the complete development of seeds. The seed coat contains three layers: sarcotesta, sclerotesta and endotesta. *Ginkgo biloba* is a shade-intolerant species growing predominantly on limestone substrates with the level of pH ranging between 5.0 and 5.5 and located at an altitude from 100 to 1500 m.

Chapter 2 - Ginkgo preparations are mainly used for the treatment of peripheral vascular disease or cerebral insufficiency, especially by elders. Supraoptimal concentration of several metals e.g., Fe, Zn, Cu, Al or Mn have toxic actions on nerve cells and neurobehavioral functioning, which can be expressed either as developmental effects or as an increased risk of neurodegenerative diseases in old age. Element content and composition of ginkgo leaves are important to know exactly by using whole leaves as drug in tea preparations or tinctures.

Macro and micro elements of ginkgo (*Ginkgo biloba* L.) leaves collected from three different sites in Hungary (Europe) were analysed. Leaves were collected from male and female trees of similar age and grown next to each other. Leaf collection took place five times in the vegetation period (from May to September) of three concecutive years. Leaves were dried at 30°C, and pulverized directly before the analysis. Element composition was determined by ICP-OES.

Well detectable differences were found among the collection sites in the element concentrations of Al, Ca, Fe, K, Mg, Mn, Na, P, Sr, Zn. However, no significant differences could be detected in the element composition of ginkgo leaves over the three-year period examined. Accumulation dynamism of the elements was quite diverse during the vegetation period. Concentration of the most elements increased continuously in the leaves. Concentrations of Al, B and Ca increased rapidly at the end of summer. During the vegetation period, concentration of Al, Ca, Mg increased by 1,5-2-fold, since content of B and Ba was more than 3-fold higher at the end of the investigation.

Element compositions of male and female trees were compared in every sampling time, eliminating the effect of phenophases. Significant differences were obtained between the sexes for some elements. Concentrations of Al, Fe, K, Na, P, Zn were higher in leaves of female trees than those in leaves of male trees. However, concentration of Ca seemed to be higher in the leaves of male ginkgo trees.

Chapter 3 - *Ginkgo biloba* L. has been used as a traditional Chinese herbal medicine for thousands of years. Several studies have already appeared on the ginkgo tree and its chemical constituents. As a result of the intense pharmacological and clinical research, phytopharmaceuticals based on partially purified ginkgo leaf extracts are now among the most sold drugs all over the world. Although many different components contribute to the overall pharmacological effect of ginkgo extracts, ginkgolides are considered to be responsible for a significant part of the beneficial effects. Standardized ginkgo leaf extracts (EGb), containing 24% flavonol glycosides and 6% terpene trilactones, are primarily prescribed for problems associated with a poor central and peripheral blood circulation like dementia, vertigo and tinnitus. Ginkgo is reported to be efficient for the treatment of Alzheimer's disease and cardiovascular disease.

Allergenic and toxic compounds, such as ginkgotoxin and ginkgolic acids were also found in ginkgo leaves. These constituents can be present in ginkgo leaf teas, while they are removed from products containing the standardized leaf extract. Recently dried ginkgo leaves are being commonly consumed on a daily basis for their stimulant properties.

Ginkgo products are sold mainly not as medicines but as dietary supplements due to the less strict regulation of herbal products considering purity and potential. The pharmaceutical quality of the different ginkgo preparations is highly depending on the chemical composition of the ginkgo leaves, the extraction method and the dissolution rate. These all determine the bioavailability and efficacy of the different products.

Chapter 4 - This chapter presents the results of a screening of different crude extracts, infusions and standardized extract from the *Ginkgo biloba* (Ginkgoaceae) leaves for total phenolic content, concentration of flavonoids and *in vitro* antioxidant activity. Main reason for this study is the determination of these parameters and their variability among the plant extracts obtained by different solvents and water infusions prepared using different methods respectively, as well as standardized extract. Results for total phenolic content determined using Folin-Ciocalteu reagent and expressed in term of gallic acid equivalent, GAE (mg of GA/g of extract) ranged from 27.47 ± 0.29 to 141.60 ± 0.36 mg of GA/g of extract and plant material. The concentrations of flavonoids determined using spectrophotometric method with aluminum chloride and expressed in terms of rutin equivalent, RuE (mg of Ru/g of extract) ranged from 14.46 ± 0.23 to 231.15 ± 0.17 mg of Ru/g of extract and plant material. Obtained results for antioxidant activity of *Ginkgo biloba* extract and infusions ranged from 1408. 96 ± 2.01 to 49.75 ± 0.90 μg/ml. Results for antioxidant activity and the amount of total phenolics content and flavonoids varied according to the type of solvent used, as well as the method of preparation of the extracts and infusion. In addition, it can be concluded that there is a relation between the quantity of phenolics, flavonoids and antioxidant activity. Great variability of the studied parameters was observed comparing the effectiveness of the used solvents. The ethanolic extracts, infusion, infusion prepared with boiled water, as well as ethanolic solution of standardized *Ginkgo biloba* extract contain the greatest concentrations of phenolics compounds, especially flavonoids and showed high antioxidant activity. According to the authors' research, leaves from *Ginkgo biloba* are rich sources of phenolic compounds with strong antioxidant activity.

In: *Ginkgo biloba*
Editor: Emmett Fisher
ISBN: 978-1-63484-460-4

Chapter 1

BIOLOGY AND ECOLOGY OF *GINKGO BILOBA* L. (GINKGOACEAE)

Milan S. Stanković*
Department of Biology and Ecology, Faculty of Science,
University of Kragujevac, Kragujevac, Republic of Serbia

ABSTRACT

In this chapter, the basic biological description, phylogeographic history, as well as ecological characteristics of *Ginkgo biloba* are presented. One of the four divisions within Gimnospermae is the division Ginkgophyta. Ginkgophyta includes only the class Ginkgopsida, with the order Ginkgoales, which, in addition to four extinct families, encompasses a recent family Ginkgoaceae. The Ginkgoaceae family includes only one recent genus – *Ginkgo,* with the species *Ginkgo biloba*. The diversification of this genus began during the Jurassic and the early Cretaceous period. The first extinction of this species started in the polar regions at the end of the Miocene. Some parts of China have acquired the status of refugia for a "living fossil species" *Ginkgo biloba.* More scientific and common names exist for this species; among others, Carl Linnaeus constructed its Latin name as *Ginkgo biloba*. The height of *Ginkgo biloba* tree ranges between 20 and 30 m, with a diameter from 50 cm to 1 m or more. Shoots are differentiated into two distinct morphological types: long shoots with widely spaced leaves that subtend axillary buds; and short, or spur, shoots with clustered leaves that lack

* E-mail address: mstankovic@kg.ac.rs.

both internodes and axillary buds. The leaves have open dichotomous venation and are wedge-shaped with a semicircular undulate margin and a middle notch that divides the leaf surface into two halves. *Ginkgo biloba* is a dioecious plant. The male reproductive structures – microstrobiles are catkin-like, pollen-bearing cones, while the female reproductive structures – macrostrobiles, consist of a stalk-like peduncle and a pair of ovules. In the ovule begins the four-month long gametophyte development process, which results in the formation of two multiflagellated motile spermatozoids. After the egg cell formation, one of the motile spermatozoids carries out the process of fertilization, after which it performs the complete development of seeds. The seed coat contains three layers: sarcotesta, sclerotesta and endotesta. *Ginkgo biloba* is a shade-intolerant species growing predominantly on limestone substrates with the level of pH ranging between 5.0 and 5.5 and located at an altitude from 100 to 1500 m.

Keywords: *Ginkgo biloba*, Ginkgoaceae, Mesozoic endemorelict, "living fossil species"

INTRODUCTION

Ginkgo biloba (Ginkgoaceae) – ginkgo, ginko, silver fruit, silver peach or maidenhair tree is a plant known in science and practice for several reasons. It represents both an endemorelict species and a famous plant in folk and modern medicine. A number of specific characteristics distinguish this species in the field of systematics, evolution, phylogeography, as well as in terms of morphology, anatomy and reproduction. The first fossil finds indicate its origin from the Mesozoic – early Jurassic, approximately 180 mya, while the related extinct species originate in the Paleozoic – approximately 280 million years ago. On this basis, *Ginkgo biloba* is described as a "living fossil". *Ginkgo biloba* is witness much of the history of plants and animals, movement of the Earth's continents, as well as of the evolution of Man. The *Ginkgo biloba* history of distribution is divided into two periods; namely after a circumpolar distribution in the Northern Hemisphere, the areal was reduced to a few populations on the refugial habitats in the territory of China, where this species survived into the Pliocene and Pleistocene epochs. Some morphological characteristics and the specificity of the reproductive process indicate its evolutionary primitivity. This species is characterized by a series of archaic biological traits, like leaf shape, structure, as well as dichotomous venation of leaves, differentiation of shoots, reproduction, multiflagellated motile spermatozoids, lack of dormancy. *Ginkgo biloba* is a species resistant to different biological, chemical and physical stress. Many attacks of phytopathogenic parasites and pests as well as pollution and nuclear radiation do not diminish the vitality of the species in question which is why it is regarded to be higly convenient for cultivation in city parks or for tree alleys on busy city streets. Some specimens that have remained intact after the nuclear bombing of Hiroshima, prove the resistance of this plant species to harmful impacts [1-6].

Due to the religious cults of *Ginkgo biloba* in Buddhism and Confucianism, this plant has been cultivated in China, Japan and Korea for a very long time. This is proven by the existence of more than 1000 years old *Ginkgo biloba* trees located in the temple courtyards. Worldwide anthropogenic distribution of this species began 300 years ago. Consequently, *Ginkgo biloba* is an object of research in many areas of biological science. The initial use in ancient Chinese medicine has been extended to the planetary use. *Ginkgo biloba* is well-known as edible and medicinal plant. It represents a source of bioactive compounds with different biological and therapeutic effects [4, 7, 8].

This chapter contains the biology and ecology of *Ginkgo biloba*, as well as the description of its evolutionary history. The section about biological characteristics elaborates on its taxonomic and systematic status, the phylogeny of the family and the genus of this plant species, as well as the nomenclature and the origin of the scientific name and the common names from an aspect of etymology. In the next part of the chapter about biology, the habitus of this plant species is described in detail (root, stem, branching, shape and venation of leaves, strobilus and seeds), as well as the reproduction process and specificity of growth and development. A representation of ecological characteristics of *Ginkgo biloba* encompasses endemic and relict status of this plant, biogeographical history of distribution, including a description of natural habitats, from the aspects of temperature, water, and light regimes and dispersion of seeds – zoochory.

SYSTEMATICS AND EVOLUTIONARY HISTORY

Seed plants – Spermatophytes (Phanerogams or Phenogamae), include Gimnospermae with divisions (phylums) Cycadophyta, Ginkgophyta, Coniferophyta and Gnetophyta, as well as Angiospermae (flowering plants) with Monocotiledones and Dicotiledones (Figure 1). The main characteristic of Gimnospermae (gymnos: naked + sperm: seed) is the absence of carpel layer around the ovule at the time of pollination, because ovules are located on megasporophylls. The other characteristic is that Gimnospermae are nonflowering plants [9]. In addition to well-known conifers, other taxa within Gimnospermae are characterized by a large number of specificities, from biological characteristics to diversity and distribution.

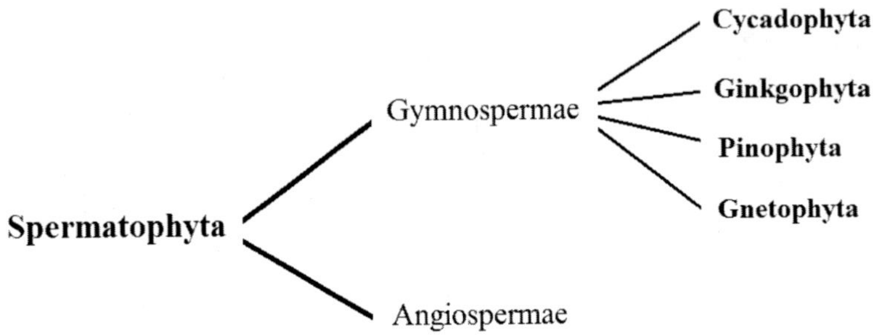

Figure 1. Phylogenetic relationships of Spermatophyta taxons.

One of the four divisions within Gimnospermae is the division Ginkgophyta. Ginkgophyta represents a significant division, particularly with respect to its extinct and recent taxa. As it has been significantly reduced in terms of quantity in recent times, monotype groups became the corresponding lower taxa (higher taxon includes only one next lower taxon). Lobed leaves with dichotomous venation, dioecy, motile sperm, long and short shoots, seeds with fleshy seedcoat (sarcotesta), etc., are the main biological characteristics that distinguish extinct and recent representatives. Based on the fossil record, the history of the Ginkgophyta taxons began in the early Mesozoic Era, although the earliest evidence was found in the Permian (the last Period of Paleozoic Era). The rise of diversification took place during the Jurassic with peaks during the Cretaceous. The origin and evolutionary-phylogenetic history of this taxon are not thoroughly understood. It is noticeable at first glance that Ginkgophyta do not have morphologically related taxa. The specificity of morphology and mode of reproduction is reflected back in an attempt classification of taxa, with insufficiently specified ancestors and relatives. Based on this inference, this taxon is classified into the division (phylum) Ginkgophyta, which together with phyla Cycadophyta, Pinophyta and Gnetophyta comprises the Gymnospermae group.

The complex multilevel analysis which is based on the fossil record of the earliest origins from the Paleozoic, their morphology, the reconstruction of habitus, the structure of reproductive organs and on the recent molecular studies, confirms the current phylogenetic position of this taxon. Additionally, some important characteristics with a systematic value indicate similarity, whereas some features distinguish Ginkgophyta from the other Gymnospermae groups. The structure of the reproductive organs, the characteristics of gametophyte and motile sperm cells indicate similarities with Cycadopsida, while the substantial difference is reflected in habitus.

The cross section anatomy of this species bears similarities to the cross section anatomy of Conifers, but the structure of reproductive organs and multiflagellated sperm cells confirms a significant difference. The modern genome analysis indicates that the difference between Ginkgophyta and Conifers is much greater in relation to the difference between Ginkgophyta and Cycadophyta, which confirms a degree of phylogenetic relatedness [4]. Taking account of what was previously proven, it is determined that the group Pteridospermatophyta (seed ferns), and most probably its order Peltaspermales, represents the ancestor of the order Ginkgoales [10].

The division Ginkgophyta includes only the class Ginkgopsida, with only one order – Ginkgoales, which, furthermore, encompasses a recent family Ginkgoaceae, in addition to four (Karkeniaceae, Umaltolepidiaceae, Yimaiaceae and Schmeissneriaceae) extinct families (Figure 2). The phylogenetic relationships within the class Ginkgopsida and both the evolution and phylogeny of the order Ginkgoales are very questionable. The main reasons are the heterogeneity of the fossil record and the leaf heterophylly within the same species, which cause incomplete classification of some lower fossil taxa within the group. However, the taxonomic significance of ovulate organs largely enables the phylogenetic reconstruction and the positioning of this taxon [5].

Ginkgoaceae family includes only one recent genus – *Ginkgo* and nine extinct genera (*Baieroxylon*, *Cheirophyllum*, *Chiropteris*, *Ginkgoites*, *Ginkgoïdium*, *Ginkgopitys*, *Phoenicpsis*, *Polyspermophyllum* and *Trichopitys*). Based on current paleobotanical research, the genus *Ginkgo* is differentiated into eleven species among which *Ginkgo biloba* is recent and the others (*Ginkgo adiantoides*, *Ginkgo apodes*, *Ginkgo cranei*, *Ginkgo digitata*, *Ginkgo dissecta*, *Ginkgo gardneri*, *Ginkgo ginkgoidea*, *Ginkgo huolinhensis*, *Ginkgo huttonii* and *Ginkgo yimaensis*) are extinct fossil species (Figure 2 and 3).

The fossil remains of the vascular system and leaves from the oldest Ginkgophyta representatives dating back to the early Permian (approximately 298-252 million years ago) period – Paleozoic Era [11], are not systematically verified due to the absence of the associated reproductive organs. Over a period of diversification, a larger number of Ginkgopyta species were widely distributed, mostly in the Northern Hemisphere. During the long period of existence, the order Ginkgoales consisted of the species from both existent Ginkgoaceae family and other, extinct families: Karkeniaceae, Umaltolepidiaceae, Yimaiaceae and Schmeissneriaceae [12].

Division	Class	Order	Family	Genus	Species
Ginkgophyta	**Ginkgopsida**	**Ginkgoales**	**Ginkgoaceae**	***Ginkgo***	***Ginkgo biloba***
					† *Ginkgo adiantoides*
					† *Ginkgo apodes*
					† *Ginkgo cranei*
					† *Ginkgo digitata*
					† *Ginkgo dissecta*
					† *Ginkgo gardneri*
					† *Ginkgo ginkgoidea*
					† *Ginkgo huolinhensis*
					† *Ginkgo huttonii*
					† *Ginkgo yimaensis*
				†*Baieroxylon*	
				†*Cheirophyllum*	
				†*Chiropteris*	
				†*Ginkgoites*	
				†*Ginkgoïdium*	
				†*Ginkgopitys*	
				†*Phoenicpsis*	
				†*Polyspermophyllum*	
				†*Trichopitys*	
			†Karkeniaceae		
			†Umaltolepidiaceae		
			†Yimaiaceae		
			†Schmeissneriaceae		

Figure 2. Systematics of Ginkgophyta, recent and extinct taxa are included († - extinct taxons).

With the advent of the first fossil record of the reproductive organs from the Early Jurassic (180 million years), the presence of representatives from the *Ginkgo* genus was definitely confirmed.

Apart from *Ginkgo* species, the family Ginkgoaceae encompassed a total of ten genera in its history [3].

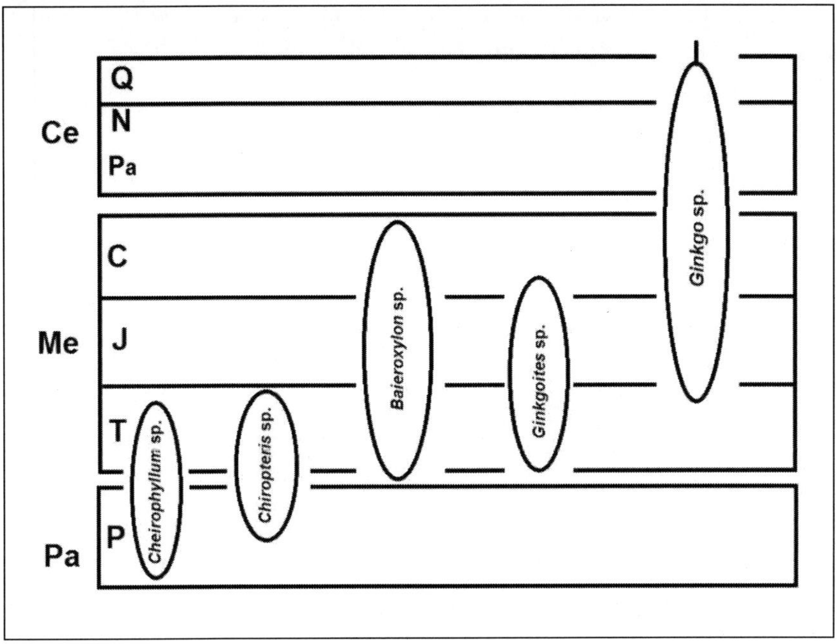

Figure 3. Temporal distribution of the existance of *Ginkgo* and related genera (P – Paleozoic; M – Mesozoic; C – Cenozoic; P – Permian; T – Triassic; J – Jurassic; C – Cretaceous; Pa – Paleogene; N – Neogene; Q – Quaternary).

The first representative of the *Ginkgo* genus described on the basis of the fossil findings from Henan – China is extinct *Ginkgo yimaensis* (Figure 4) with a distinctly notched leaves and small ovules, very different in comparision with recent *Ginkgo biloba* and related species [2, 13]. Another described fossil species from this genus is *Ginkgo adiantoides*, morphologically (due to the morphology of leaves and generative organs) very similar to *Ginkgo biloba*. The similarities indicate the close affinity between these two species. *Ginkgo adiantoides* existed from the Cretaceous until the end of the Tertiary (now the Pliocene in the Neogene). The species from the *Ginkgo* genus and other genera of the family *Ginkgoaceae* were very numerous as forest communities across the Northern Hemisphere [2].

The diversification of this genus began during the Jurassic and the early Cretaceous period in the Mesozoic Era, mainly in the area of the Northern Hemisphere. However, bearing in mind the position of the Earth's continents at the beginning of the diversification, it is inferred that it was the territory of Laurasia. The fossils discovered all over the world prove this fact, but there is no data about the distribution in the Equatorial regions [7]. The period of intense and broad diversification (120 million years ago) is the end of the Mesozoic (the Cretaceous) and the beginning of the Cenozoic (the early Paleogene) Era [14].

Figure 4. *Ginkgo yimaensis*: the appearance of a short shoot with macrostrobiles and deeply divided leaves – reconstruction based on fossil data.

The first early extinction was noticed among the fossil record of the late Cretaceous (98-65 million years ago). These fossil remains provided evidence of the extinction of the species *Ginkgo huolinhensis*, as well as of the reduction of diversity and distribution of other species of this genus [15]. Furthermore, the fossil remains indicate that during the Palaeocene, in the territory of the Northern Hemisphere, just a few other species, such as *Ginkgo cranei* and *Ginkgo adiantoides*, existed. The extinction of the species continues during the Tertiary (now the Paleogene and the Neogene), where the significant progress of extinction was observed during the Oligocene (38-26 million years ago). The last fossil remains of the *Ginkgo* species were found at the end of the Pliocene (5.3-2.5 million years ago). During this period the areal of *Ginkgo biloba* reduced drastically in size so that it now covers only a small part of the territory of China. All the fossils from the Tertiary and from the early Pleistocene were found in East Asia, on the teritory covered with the current *Ginkgo* areal [16]. The last known fossil remains from the period of the Pleistocene were found on the southwestern territory of Japan. The first extinction of the *Ginkgo biloba* species in the northern areas commenced during the Miocene. Moreover, it is proven that the species died out on the territory of North America at the end of the Miocene Epoch, whereas the utter extinction on the Continent occurred at the end of the Pliocene [17].

The main reason for the extinction and the reduction of the number of Ginkgophyta species is the drastic annual temperature decrease that occurred prior to the Ice Age. The sensitivity of these species is regarded to be the cause of their early extinction. This is confirmed in the studies proving that not a single species of *Ginkgo* genus existed during the Pliocene in Europe. The climate regime changes, particularly the temperature drop with polar and alpine ice progression in the Pleistocene era, caused alterations in the diversity and distribution of wildlife, primarily in the northern continental parts such as North America, northern Europe and extremely severe parts of Asia [18]. The important floristic changes are the result of historical moments such as the withdrawal of species due to the spread of the ice sheets, temperature fluctuations (glacial and interglacial periods), and in the end, the recolonization of species during the postglacial period. Certain species were forced to extinction at the beginning, during and after the ending of the glacial period due to their either partial or total resistance to temperature variations [5]. The spatio-temporal dynamics of the Ice Age was with some variations in the Western Asia. Namely, certain areas such as parts of the territory of China, were protected from the direct influence of the Ice Age, and have thus gained the status of refugia [19-22].

A number of the Tertiary species have been successfully preserved from extinction in refugial habitats [23]. Their main characteristic is offset by the effect of the Pleistocene Ice Age.

The resistance to temperature fluctuations and the preservation of both species and habitats in refugia on the territory of China are the main reasons for the current existence of a "living fossil" *Ginkgo biloba* [3]. In a review of the chorology dynamics, the general life history of the Ginkgophyta species, from the Permian to the present day, is very different. For some species, the Pliocene and the close previous periods are the periods of the total disappearance, but for *Ginkgo biloba* these were the periods of sharp reduction in both distribution and habitation in glacial refugia [3, 24].

Specific historical moments and the biology of *Ginkgo biloba* are the reasons for this species with both endemorelict status (Mesozoic relict) and status of a witness of the seed plant and human evolution to survive to the present day [5, 25]. Based on this, *Ginkgo biloba* was included in the IUCN Red List of Threatened Species.

Because they provide a survival of this plant, the refugial habitats of *Ginkgo biloba* in China are widely studied from different aspects, such as molecular ecology, phytosociology, vegetation history, paleobotany, conservation ecology as well as anthropogenic distribution of *Ginkgo biloba* etc. In order to clarify the distribution history of *Ginkgo billoba,* genetic comparisons with a larger number of samples taken from different locations were made [3].

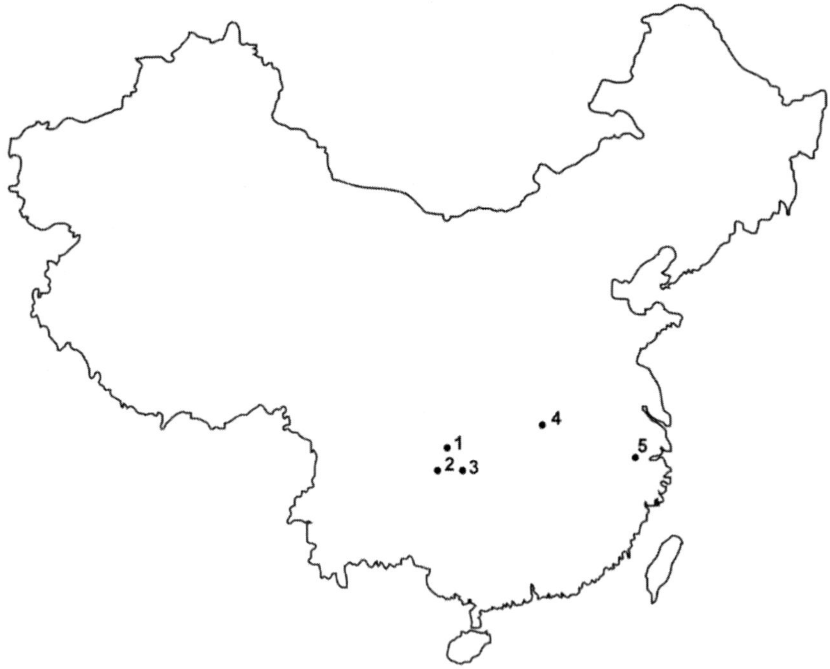

Figure 5. Positions of refugial habitats of *Ginkgo biloba* in the territory China, (1 – Mt. Jinfo, Chongqing; 2 – Mt. Dalou, Guizhou and Sichuan; 3 – Wuchuan, Guizhou; 4 – Mt. Dahong, Hubei; 5 – Mt. Tianmu, Zhejing).

According to the numerous biological and environmental studies, the natural sites of *Ginkgo biloba*, proven as refugia of these Mesozoic endemorelict species in Southwest China (Figure 5) are: Mt. Jinfo – Chongqing, Mt. Dalou – Guizhou and Sichuan, Wuchuan – Guizhou, Mt. Dahong – Hubei, Mt. Tianmu – Zhejing [6, 8, 26-34]. The recent genetic research pointed out the questionable status of the natural origin of some glacial refugia [8].

ETYMOLOGY

The existence of diverse scientific and popular names for *Ginkgo biloba* indicates the interest for this species in science and practice. German physicist and botanist Engelbert Kaempfer (1651-1716) was the first to publish a description and the name of this species. During his two-year stay (1690-1692) in Japan, Kaempfer dealt with medicinal plants of Japan ("*Flora Japonica*", published as part of the "*Amoenitatum exoticarum*", Lemgo 1712), and with history of Japan ("*History of Japan*", 1727). Ever since that time, *Ginkgo biloba* was very well-known and respected in Japanese religion. Kaempfer first encountered this plant during a visit to Nagasaki Buddhist temple, at the beginning of the 1691. Then he recorded the description of this plant with the name *Ginkgo* as the first European in the history of the nomenclature of this species to do so. Below is his original Latin text and illustration as published in this work (5th volume, page 811-813). During the visit, he brought a few seeds of *Ginkgo biloba* and planted them in the Utrecht botanical garden, thus making it the first exemplar transferred from the East to Europe [35]. In 1771, Carl Linnaeus (1707-1778) constructed the Latin name based on the principles of binomial nomenclature with the old name "Ginkgo" for genus name and an epithet "biloba" (Figure 6) as a description of leaf (Latin: bis, bi – two + loba, lobed). *Ginkgo biloba* L. is, therefore, the official name in botanical literature [36].

On the basis of the description from British botanist Richard Anthony Salisbury (1761-1829), James Edward Smith (1759-1828) constructed the name of this species in 1979 as *Salisburia adiantifolia* Smith (due to the similarities with fern leaf from genus *Adiantum* – maidenhair ferns). John Gudgeon Nelson named the species in 1866 (1818-1882) as *Pterophyllus salisburiensis* J. Nelson. Even though both names are synonymous neither one is in official use [37, 38]. In analyzing the life history and relict traits, Charles Darwin (1809-1882) called this species "a living fossil" [1]. The common name is different in various countries: in English, the name is most similar to the botanical name – Ginkgo tree or Maidenhair Tree, in Chinese: bái guǒ – white fruit, yínxìng – silver apricot, yínguǒ – silver fruit, in Japanese: ginnan, in Korean: eunhaeng, in Polish: Miłorząb dwuklapowy, in German: Älterer Ginkgobaum, in Russian: Гинкго, in Serbian: Гинко, in Spanish: albaricoque plateado, and in French: L'Arbre aux quarante écus or l'Abricotier d'argent etc. Variant spellings include ginko and gingo.

Figure 6. Typical appearance of *Ginkgo biloba* leaf.

MORPHOLOGY

The height of *Ginkgo biloba* tree is in the range from 20 to 30 m, whereas in the case of some very old specimens, the height varies between 40 and 50m. The crowne form varies depending on plant age: the crown of very young plants (Figure 7) is unintegrated and moderately branched, and is later to acquire a conical shape, while old specimens form a moderately oval crown, with a well integrated structure. The main branch of the central tree is branched depending on the light regime (Figure 8 and 9). The diameter of the tree also varies depending on the age, from 50 cm to 1 m, and in the case of some very old individuals the diameter ranges between 3 and 5 m. *Ginkgo biloba* individuals are very well rooted with a well-branched mycorrhizal root system, which provides exceptional stability on highly degraded and steeper habitats. A stable ground part and a strong stable canopy combined with a strong root allow individuals good resistance to winds and retention of snow during winter. Due to the exceptional stability of habitats and the resistance of the species to various parasitic attacks, *Ginkgo biloba* can have a long life span of 1.000 or 2.000 years.

Figure 7. Ginkgo biloba – a young plant.

Figure 8. Branching of the central tree.

Older *Ginkgo biloba* individuals are characterized by the phenomena of sprouting from embedded buds in the basal part of the trunk. This process of sprouting results in the development of lignotubers. In some cases, several lignotubers can be differentiated from the main trunk. Lignotubers have positive geotropic orientation. After the initial phase of thickening of the main trunk, lignotubers can be observed below the surface of the substrate. Apart from lignotubers, there is another type of specific vegetative formation called aerial root. Aerial roots develop at a certain height from the substrate, growing positively geotropic to the surface of the substrate. When it comes to older individuals, aerial roots support large lateral branches. Upon touching the ground surface, these roots form lateral shoots.

Figure 9. *Ginkgo biloba* – a mature plant.

Figure 10. Surface of *Ginkgo biloba* stem bark.

The wood structure of *Ginkgo biloba* is similar to the structure of coniferous trees. Older individuals have a thickened and cracked stem bark (Figure 10). Regardless of a certain similarity to conifer species, one of the inherent differences is the impossibility of resin formation due to the absence of resin canals. *Ginkgo biloba* shoots can be differentiated into two distinct morphological types: long shoots with widely spaced leaves that subtend axillary buds and short shoots, or spurs, with clustered leaves that lack both internodes and axillary buds (Figure 11). Long shoots are responsible for the development of the basic framework of the tree and generation of new growing points, while the short shoots produce the majority of leaves and reproductive structures [2].

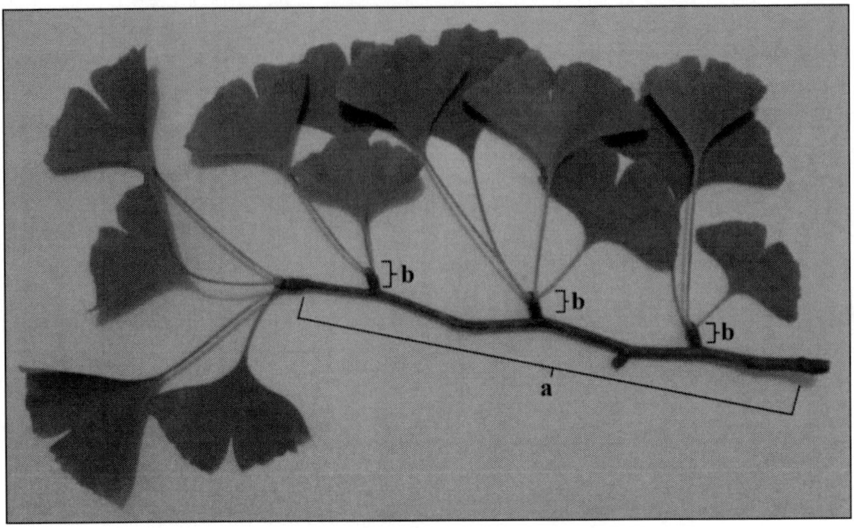

Figure 11. *Ginkgo biloba* long shoot (a) with short shoots (b).

The comparative histological analysis confirmed the difference in structure between these two groups of *Ginkgo biloba* shoots. The shoot differentiation enables both characteristic branch formation and even and dense leaf distribution on the branches. External changes, such as injuries, can induce a transformation of short shoots. The formation of long and short shoots was first recorded in the fossil *Ginkgo* species. *Ginkgo biloba* is a deciduous plant. Its leaves develop in spring from buds that are well protected with bud scales. In autumn, in a very short period of time, the leaves become yellow and drop.

Apart from numerous distinctive features, the leaf structure and shape represent important specific characteristics of this species. The leaves are wedge-shaped with a semicircular undulate margin and a middle notch, of a variable length, that divides the leaf surface into two halves. The length of the petiole varies from 5 to 10 cm, whereas the whole leaf length ranges from 5 to 15 cm. The petiole is thin and very flexible. A specific leaf form is the reason for the epithet in the binomial species name: "biloba" (Latin: bis, bi – two + loba, lobed).

The differentiation of shoots of *Ginkgo biloba* induces the differentiation of the leaf form. The leaves that develop on long shoots have a leaf central notch of different length, while the leaves on short shoots have no notch. The comparative analysis of the fossil remains suggests that the leaves of *Ginkgo* species tended to increase the integrity of the leaf surface, from the species with more highly dissected leaves such as *Ginkgo yimaensis* to the species with the entire leaf surface or with just one notch such as *Ginkgo adiantoides* or *Ginkgo biloba*. In addition to the size variability – anisophylly, there is high variability in shape – heterophylly (Figure 12).

Figure 12. Heterophily of *Ginkgo biloba* leaves.

Leaf heterophylly is conditioned by several factors and can occur in several different forms. Apart from the type of shoots, plant age has an impact on the level of leaf shape variability. Based on these findings, the shoots of a very young individuals develop leaves with a few notches, similar to the leaves of ancestral species, thus representing an example of the theory: ontogeny recapitulates phylogeny. In addition to the plant age and type of shoots, ecological conditions, mostly light, induce leaf heterophylly [39]. The leaves are hypostomatic – most of the stomata are located on the abaxial leaf side, while the rest is located on the adaxial side. Another peculiarity of the *Ginkgo biloba* leaf is reflected in the open and dichotomous type of venation (Figure 13).

Figure 13. Dichotomy venation of *Ginkgo biloba* leaf.

REPRODUCTION

Ginkgo biloba is a dioecious plant with male and female organs forming on different plants. Some 20 to 30 years are necessary for this species to achieve the full maturity. During this period, the gender of a specimen is not known. The ratio of male and female individuals in the wild population varies due to the impacts of several factors: in healthy undisturbed populations ratio is 1:1 and can vary up to 3:2. Simultaneously with the development of leaves, male and female reproductive organs or micro- and macrostrobiles develop from short shoots. As is the case with anemophilous plants, *Ginkgo biloba* is effectively pollinated prior to the full leaf development. The maximum distance of pollen distribution is not exactly defined, but bearing in mind the characteristics of anemophilous plant species, it is possible that it is a great distance. This possibility is even more justified with the fact that some female plants are pollinated without the existence of very close male plants.

The male reproductive structures – microstrobiles (Figure 14), develop on short shoots of male plants. Microstrobiles are catkin-like, pollen-bearing cones, with a flexible axis, from 5 to 10 cm long, and with stalk-like microsporophylls, each of which bears an apical pair of elongate, pendulous microsporangia.

After the opening of microsporangia, the mature pollen grains are blown by wind (i.e., anemophily). When the process of the pollination is completed, the microstrobiles fall off, while the leaves continue to develop. The ripening and the opening of microsporangia and consequently, the dissipation of pollen grains depend on the climatic conditions and occur most commonly in the period from the beginning of April to the end of May. One short shoot can develop several microstrobiles, most often from 3 to 10, each of which produces from 30 to 50 microsporangia. The comparative studies of the *Ginkgo biloba* pollen morphology suggest similarities with the pollen of the *Cycas* species; however, there are differences in the structure of the pollen exine.

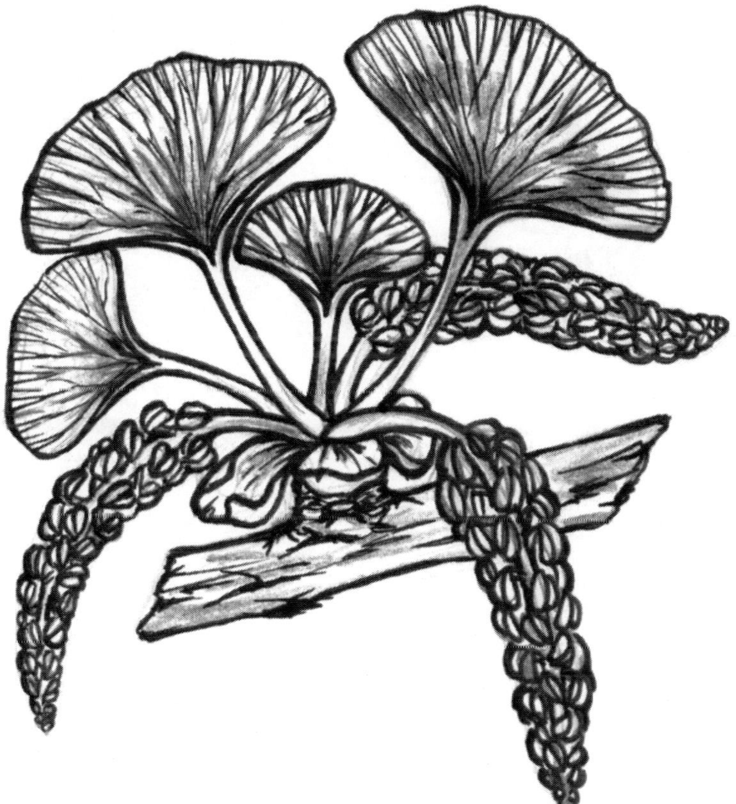

Figure 14. Short shoot of male *Ginkgo biloba* plant with pollen bearing cones.

The female reproductive structures – macrostrobiles (Figure 15), develop on short shoots of female plants. Macrostrobiles consist of a stalk-like peduncle, whose length varies from 3.5 to 4 cm, and a pair of ovules, usually each up to 2 or 3 mm long. One short shoot of female plant can develop 4-6 macrostrobiles. In most of the cases, only one ovule develops after the insemination.

Figure 15. Short shoot of female *Ginkgo biloba* plant with ovules.

However, seldom do both ovules of a pair develop simultaneously. In the case of some extinct species, there were more ovules in one peduncle. When the ovule is receptive, it secretes a small droplet of mucilaginous fluid from its micropyle whose function is to capture airborne pollen. The retraction of this droplet at the end of the day brings the pollen into the pollen chamber. In the ovule begins the four-month long gametophyte development process, which results in the formation of two multiflagellated motile spermatozoids, only one of which fertilizes a waiting egg cell. The similar characteristics of reproductive organs are noticeable in the *Cycas* species [2, 40]. The existence of a motile sperm was first proved in 1894 by Sakugoro Hirase who researched the process of fertilization on the *Ginkgo* plant in the Botanical garden of the University of Tokyo.

What follows after the formation of megaspore and divisions in the nucellus of ovule, is the creation of gametophyte and decrease of the volume of nucellus. The apex of the gametophyte forms two archegonial cells, after which the membrane of the megaspore allows the contact of the nucellus with the chamber by means of apical resorption. After the egg cell formation, a sperm fertilizes an egg cell and one of the motile spermatozoids carries out the several-months long process of fertilization. In some cases, the embryo continues its development even after the seed drops from the tree.

Figure 16. Short soot of female *Ginkgo bloba* plant with mature *Ginkgo biloba* ovules – seeds.

During the process of further development, the growth of the ovules continues until they reach from 2 to 3 cm in diameter (Figure 16). Their seed coats, differentiated during the formation of the ovules, consist of a soft, fleshy outer layer (the sarcotesta), a hard, stony middle layer (the sclerotesta), and a thin, membranous (the endotesta) inner layer (Figure 17). In the developmental phase, the ovules are green, with stomata on the surface, whereas in the postmaturation phase, they turn yellow. The *Ginkgo* mature ovule (seed) releases an unpleasant odor due to the presence of volatile compounds in sarcotesta such as butanoic and hexanoic acids. The mature seeds are distributed by animals – *Ginkgo biloba* is zoochorous species. The presence or absence of sarcotesta significantly determines the beginning of seed germination [41].

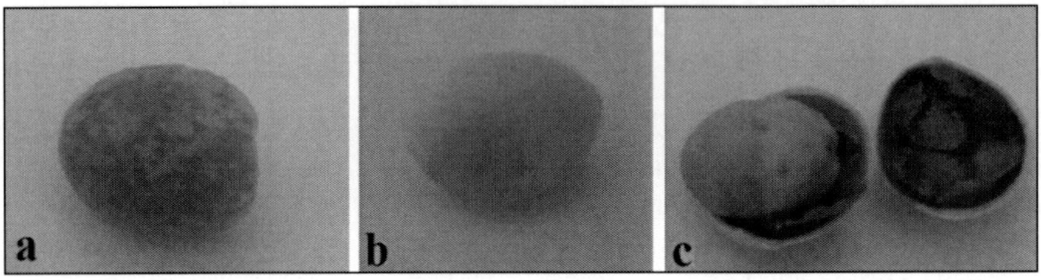

Figure 17. *Ginkgo biloba* seed with sarcotesta (a), with sarcotesta removed (b) and with sclerotesta and endotesta desintegrated (c).

Ecology

In addition to the three hundred years old wild and half-wild populations in China, the *Ginkgo biloba* species grows in many different climates, on different substrates and under different water and temperature regimes. The adaptability of this species enables its world-wide cultivation. Furthermore, the adaptation of natural populations to the environmental factors can be determined by means of analysis of both a small number of natural habitats in China and of the fossil records. The analysis of ecological conditions in the habitats of natural populations in China suggests that *Ginkgo biloba* predominantly grows on limestone substrates, with pH from 5.0 to 5.5 and at an altitude from 100 to 1500 m. The data obtained from in-depth analysis of the fossil remains show riparian habitats as predominant environment of the species in question. *Ginkgo biloba* is characterized as a shade-intolerant species due to the fact that it grows in sunny habitats [6, 42]. The stable root system, the secondary and aerial roots and the resistance of the species to parasites enable its good adaptability.

In addition to growth and development, the temperature is one of the important factors that determine the dynamics of pollination and seed germination. This is clearly visible if the time required for these processes in different climates is taken into account. In warm climates, the pollination lasts from March to April, the seed abscission occurs in September and the germination takes place in March of the following year. In colder climates, the pollination happens in May, the seed abscission in October and November, while the germination takes place in June next year. Similar variability is observed at increasing altitude at which the plant grows and develops [43].

The floristic composition of the habitats in which *Ginkgo biloba* survived to the present day, may largely reflect the environmental conditions particularly favourable for this species. The flora of Dalou Mountains belongs to the Sino-Japanese floristic region, while the climate is subtropical – humid and warm. Other conditions such as the annual rainfall ranging from 1270 to 1400 mm, the average annual temperature varying from 13 to 15.3°C, with average monthly minimum from 1.8 to 4.5°C in January and a maximum from 22 to 25.1°C in July, as well as average annual humidity of 80% correspond to the subtropical monsoon area [6].

The following list of woody species was recorded on the same site and show similarities with the floristic composition of *Ginkgo biloba* fossil habitats: *Liquidambar formosana* Hance, *Cunninghamia lanceolata* (Lamb.) Hook., *Taxus wallichiana* var. *chinensis* (Pilger.) Florin, *Lindera megaphylla* Hemsl., *Cyclobalanopsis glauca* (Thunb.) Oerst., *Cornus controversa* Hemsl., *Choerospondias axillaris* var. *pubinervis* (Rehd. et Wils.) Burtt et Hill, *Juglans cathayensis* Dode, *Celtis biondii* Pamp., *Quercus aliena* Bl., *Cinnamomum wilsonii* Gamble, *Machilus nanmu* (Oliv.) Hemsl., *Sapium sebiferum* (L.) Roxb., *Acer laevigatum* Wall., *Prunus dielsiana* (Schneid.) Yü et Li, *Tilia tuan* Szyszyl., *Ilex micrococca* Maxim., *Diospyros cathayensis* Steward, *Corylus chinensis* Franch., *Osmanthus yunnanensis* (Franch.) P. S. Green, *Michelia martini* (Lévl.) Lévl., *Salix hypoleuca* Seemen, *Mallotus apelta* (Lour.) Müell. Arg., *Ficus henryi* Warb. ex Diels, *Cupressus funebris* Endl., *Castanopsis tibetana* Hance, *Hovenia acerba* Lindl., *Aphananthe aspera* (Thunb.) Planch., *Tapiscia sinensis* Oliv. and *Emmenopterys henryi* Oliv [6].

In relation to the biotic factors, the influence of animals is very important. When it comes to this species, endozoochory represents the usual mode of seed dispersal. The juicy and nutritious sarcotesta makes *Ginkgo biloba* seeds appropriate for regular consumption by mammals. Consequently, mammals, via their ingestion, disperse seeds.

The paucity of fossilized *Ginkgo* seeds has not deterred speculation as to what animals might have dispersed seeds over the course of its long evolution [2]. Some evidence suggests that it could be some dinosaur species, the first bird species, as well as Multituberculata – the species of mammalian rodents. The studies of the *Ginkgo* seed distribution confirmed that zoochoric animals are: leopard cat (*Felis bengalensis*, Felidae) and masked palm civet (*Paguma larvata*, Viveridae) in China, and raccoon dog (*Nyctereutes procyonides*, Canidae) in Japan. Moreover, local squirrel species (family Sciuridae) can distribute seeds in the wild populations as well as in plantations.

Due to the worldwide distribution, the *Ginkgo biloba* species is confirmed to be extremely beneficial in the domains of pharmacy and medicine. Apart from medicinal application, the species found its place in the field of horticulture. Therefore, it is easily inferrable that the anthropogenic factor, as one of the biotic factors, generally has a favourable effect on this species.

CONCLUSION

Ginkgo biloba (Ginkgoaceae) – ginkgo or maidenhair tree is a plant known for many specificities in terms of biology, ecology and application. This species is the only living representative of the monotypic gymnospermous genus *Ginkgo,* as well as of monotypic family Ginkgoaceae, order Ginkgoales, class Ginkgopsida, and the division Ginkgophyta. In addition to the presented isolated phylogenetic position and ancient lineage, distinctive morphology and reproduction are the reasons for naming this Mesozoic endemorelict species as "living fossil" and registration in the IUCN Red List of Threatened Species. Widely distributed, the species represents the source of various types of application in different fields such as pharmacy, medicine and horticulture thus confirming, that in general terms, anthropogenic factor favourably influences the species.

REFERENCES

[1] Major, RT. (1967). The ginkgo, the most ancient living tree: The resistance of *Ginkgo biloba* L. to pests accounts in part for the longevity of this species. *Science,* 157 (3794), 1270-1273.

[2] Del Tredici, P. The evolution, ecology, and cultivation of *Ginkgo biloba*. In: Van Beek ART. *Ginkgo biloba*. Amsterdam, The Netherlands: Harwood Academic Publishers; 2000; 7-23.

[3] Gong, W; Chen, C; Dobeš, C; Fu, XC; Koch, AM. (2008). Phylogeography of a living fossil: Pleistocene glaciations forced *Ginkgo*

biloba L. (Ginkgoaceae) into two refuge areas in China with limited subsequent postglacial expansion. *Molecular Phylogenetics and Evolution,* 48, 1094-1105.
[4] Singh, B; Kaur, P; Gopichand, Singh, RD; Ahuja, PS. (2008). Biology and chemistry of *Ginkgo biloba. Fitoterapia,* 79 (6), 401-418.
[5] Zhou, YZ. (2009). An overview of fossil Ginkgoales. *Palaeoworld*, 18 (1), 1-22.
[6] Tang, CQ; Yang, Y; Ohsawa, M; Yi, SR; Momohara, A; Su, WH; Wang, HC; Zhang, ZY; Peng, MC; Wu, ZL. (2012). Evidence for the persistence of wild *Ginkgo biloba* (Ginkgoaceae) populations in the Dalou Mountains, southwestern China. *American Journal of Botany*, 99 (8), 1408-1414.
[7] Pang, Z; Pan, F; He, JS. (1996). *Ginkgo biloba* L.: History, current status and future prospects. *The Journal of Alternative and Complementary Medicine,* 2 (3), 359-363.
[8] Shen, L; Chen, YX; Zhang, X; Li, YY; Fu, XC; Qiu, XY. (2005). Genetic variation of *Ginkgo biloba* L. (Ginkgoaceae) based on cpDNA PCR-RFLPs: entrance of glacial refugia. Nature Publishing Group, 94 (4), 396-401.
[9] Simpson, GM. Plant systematics. 2nd edition. Academic Press; 2010.
[10] Royer, LD; Hickey, JL; Wing, LS. (2003). Ecological conservatism in the "living fossil" *Ginkgo*. *Paleontology*, 29 (1), 84-104.
[11] Zhou, ZY. (2003). Mesozoic Ginkgoaleans: phylogeny, classification and evolutionary trends. *Acta Botanica Yunnanica*, 25 (4), 377-396.
[12] Gnaedinger, S. (2011). Ginkgoalean woods from the Jurassic of Argentina: Taxonomic considerations and paleogeographic distribution. *Geobios,* 45 (2), 187-198.
[13] Zhou, Z; Zhang, B. (1989). A middle-Jurassic *Ginkgo* with ovule-bearing organs from Henan, China. *Palaeontographica,* 211 (4-6), 113-133.
[14] Traula, H. (1967). Evolutionary trends in the genus *Ginkgo*. Lethaia, 1 (1), 63-101.
[15] Dong, M; Sun, G. (2012). *Ginkgo huolinhensis* sp. nov. from the Lower Cretaceous of Huolinhe Coal Field, Inner Mongolia, China. *Acta Geologica Sinica*, 86 (1), 11-19.
[16] Zhou, Z; Quan, C; Liu, YS. (2012). Tertiary *Ginkgo* ovulate organs with associated leaves from North Dakota, U.S.A., and their evolutionary significance. *International Journal of Plant Sciences*, 173 (1), 67-80.

[17] Uemura, K. Cenozoic history of *Ginkgo* in East Asia. In: Hori T, Ridge RW, Tulecke W, Del Tredici P, Trémouillaux-Guille J, Tobe H. *Ginkgo biloba* – a global treasure: from biology to medicine. Tokyo, Japan: *Springer*; 1997; 207-221.

[18] Rapp, D. Ice Ages and interglacials – measurements, interpretation and models. 2nd edition. Chichester, United Kingdom: *Springer – Praxis Books;* 2009.

[19] Wanti, D; Quingyu, P; Xihao, W. (1980). Climatic variations in China during the Quaternary. *GeoJournal*, 4 (6), 515-524.

[20] Axelrod, DI; Ai-Shehbaz, I; Raven, PH. History of the modern flora of China. In: Zhang, A; Wu, A. (Eds.). Floristic Characteristics and Diversity of East Asian Plants. New York: *Springer;* 1996; 43-55.

[21] Yu, G; Chen, X; Ni, J; Cheddadi, R; Guiot, J; Han, H; Harrison, SP; Huang, C; Ke, M; Kong, Z; Li, S; Li, W; Liew, P; Liu, G; Liu, J; Liu, Q; Liu, KB; Prentice, IC; Qui, W; Ren, G; Song, C; Sugita, S; Sun, X; Tang, L; Van Campo, E; Xia, Y; Xu, Q; Yan, S; Yang, X; Zhao, J; Zheng, Z. (2000). Palaeovegetation of China: a pollen data-based synthesis for the mid-Holocene and last glacial maximum. *Journal of Biogeography,* 27 (3), 635-664.

[22] Zhao, P; Chen, L; Zhou, X; Gong, Y; Han, Y. (2003). Modeling the East Asian climate during the last glacial maximum. *Science in China*, 46 (10), 1060-1068.

[23] Stewart, RJ; Lister, MA; Barnes, I; Dalén, L. (2010). Refugia revisited: individualistic responses of species in space and time. *Proceedings of the Royal Society B: Biological Sciences*, 277 (1682), 661-671.

[24] Sakai, A. (1971). Freezing resistance of relicts from the Arcto-Tertiary flora. *New Phytologist,* 70 (6), 1199-1205.

[25] Zhiyan, Z. (1991). Phylogeny and evolutionary trends of Mesozoic Ginkgoales – a preliminary assessment. *Review of Palaeobotany and Palynology,* 68 (3-4), 203-216.

[26] Jiang, M; Jin, Y; Zang, H. (1990). A preliminary study on *Ginkgo biloba* in the Zhongshan region, Hubei. *Journal of Wuhan Botanical Research,* 8, 191-193.

[27] Xiang, YH; Xiang, BX. (1997). Primary report on an ancient *Ginkgo biloba* deleted community in Wuchuan County, Guizhou Province. *Guizhou Science,* 15 (4), 239-244.

[28] Xiang, YH; Xiang, Z. (1999). Ancient *Ginkgo biloba* report III. investigation on ancient *Ginkgo biloba* remnant population in Guiyang. *Guizhou Science*, 17 (3), 221-230.

[29] Li, YW; Liu, ZU; Tan, YM; Ren, MB. (1999). Studies on the *Ginkgo* at the Jingshan Mountain. *Journal of Forest Research*, 12, 197-201.

[30] Xiang, Z; Zhang, ZL; Xiang, YH. (2001). Investigation of natural *Ginkgo biloba* population on the Golden Buddha mountains of Nanchuan, Chongqing. *Guizhou Science*,19, 37-52.

[31] Ge, YU; Qiu, YU; Fu, CX. (2003) Genetic diversity of the "living fossil" *Ginkgo biloba* based on ISSR marker. *Biodiversity Science*, 11, 276-287.

[32] Xiang, H; Xu, CL; Jiang, YH. (2003). A report on *Ginkgo* resources in Panxian County, Guizhou Province: Guizhou ancient *Ginkgo* resources investigation V. *Guizhou Science*, 21, 159-174.

[33] Lin, H; Zhang, DH. (2004). Analysis for the origin of *Ginkgo* population in Tianmu Mountain. *Scientia Silvae Sinicae*, 40 (2), 28-31.

[34] Xiang, BX; Xiang, Z; Xiang, YH. (2006). Investigation of wild *Ginkgo biloba* in Wuchuan county of Guizhou, China – Guizhou ancient *Ginkgo biloba* resources investigation VII. *Guizhou Science*, 24, 56-67.

[35] Haberland, D; Engelbert K. A biography. 1st edition. London, United Kingdom: British Library Board; 1998.

[36] Simpson, DP. Cassell's latin dictionary. 5th edition. London: Cassell Ltd.; 1979.

[37] Boulger, GS. Salisbury, Richard Anthony. In: Lee S. Dictionary of national biography 50. London: *Smith, Elder and Company*; 1897.

[38] Hanelt, P. Mansfeld's encyclopedia of agricultural and horticultural crops. 1st edition. Heidelberg, Germany: *Springer-Verlag*; 2001.

[39] Critchfield, BW. (1970). Shoot growth and heterophylly in *Ginkgo biloba*. *Botanical Gazette*, 131 (2), 150-162.

[40] Jin, B; Zhang, L; Lu, Y; Wang, D; Jiang, XX; Zhang, M; Wang, L. (2012). The mechanism of pollination drop withdrawal in *Ginkgo biloba* L. *BMC Plant Biology*, 12, 59.

[41] Holt, FB; Rothwell, WG. (1997). Is *Ginkgo biloba* (Ginkgoaceae) really an oviparous plant? *American Journal of Botany*, 84 (6), 870-872.

[42] Fu, L; Li, N; Mill, RD. *Ginkgo biloba*. In: Wu ZY; Raven PH; Hong DY. Flora of China IV. Beijing: Missouri Botanical Garden Press; 1999; 8.

[43] Del Tredici, P. (2007). The phenology of sexual reproduction in *Ginkgo biloba*: ecological and evolutionary implications. *The Botanical Review*, 73 (4), 267-278.

In: *Ginkgo biloba*
Editor: Emmett Fisher

ISBN: 978-1-63484-460-4

Chapter 2

ELEMENT COMPOSITION OF *GINKGO BILOBA* L. LEAVES

***Noémi Koczka*[1,*] *and Éva Stefanovits-Bányai*[2]**
[1]Institute of Horticulture, Szent István University, Gödöllő, Hungary
[2]Department of Applied Chemistry, Corvinus University of Budapest, Budapest, Hungary

ABSTRACT

Ginkgo preparations are mainly used for the treatment of peripheral vascular disease or cerebral insufficiency, especially by elders. Supraoptimal concentration of several metals e.g., Fe, Zn, Cu, Al or Mn have toxic actions on nerve cells and neurobehavioral functioning, which can be expressed either as developmental effects or as an increased risk of neurodegenerative diseases in old age. Element content and composition of ginkgo leaves are important to know exactly by using whole leaves as drug in tea preparations or tinctures.

Macro and micro elements of ginkgo (*Ginkgo biloba* L.) leaves collected from three different sites in Hungary (Europe) were analysed. Leaves were collected from male and female trees of similar age and grown next to each other. Leaf collection took place five times in the vegetation period (from May to September) of three concecutive years. Leaves were dried at 30°C, and pulverized directly before the analysis. Element composition was determined by ICP-OES.

[*] Corresponding author: E-mail: koczka.noemi@mkk.szie.hu.

Well detectable differences were found among the collection sites in the element concentrations of Al, Ca, Fe, K, Mg, Mn, Na, P, Sr, Zn. However, no significant differences could be detected in the element composition of ginkgo leaves over the three-year period examined. Accumulation dynamism of the elements was quite diverse during the vegetation period. Concentration of the most elements increased continuously in the leaves. Concentrations of Al, B and Ca increased rapidly at the end of summer. During the vegetation period, concentration of Al, Ca, Mg increased by 1,5-2-fold, since content of B and Ba was more than 3-fold higher at the end of the investigation.

Element compositions of male and female trees were compared in every sampling time, eliminating the effect of phenophases. Significant differences were obtained between the sexes for some elements. Concentrations of Al, Fe, K, Na, P, Zn were higher in leaves of female trees than those in leaves of male trees. However, concentration of Ca seemed to be higher in the leaves of male ginkgo trees.

INTRODUCTION

Mineral elements in plants become important when their health benefits are considered in the body of organism. In human diet, calcium (Ca) and phosphorus (P) assist in bones and teeth development, calcium is also vital for nerve transmission and muscle function. Magnesium (Mg) plays a role in a wide range of fundamental cellular reactions, it is an essential mineral for energy activity, like calcium and phosphorus plays a role in regulating the acid alkaline balance in the body. In addition, magnesium acts as a physiological regulator of membrane stability and in neuromuscular, cardiovascular, immune, and hormonal function. The body needs a small amount of sodium (Na) to help maintain normal blood pressure and normal function of muscles and nerves. Potassium (K) has an important role in the synthesis of amino acids and proteins, it is necessary for electrolyte balance, and controls high pressure [1] Aluminum (Al) is the most commonly occurring metallic element, and it is a major component of almost all common inorganic soil particles. Soluble aluminum is associated with the uptake and bioaccumulation of Al from soils into plants [2]. Iron (Fe) is known for haem formation, manganese (Mn) and copper (Cu) aid iron absorption in the body. Strontium (Sr) promotes calcium uptake into bones, zinc (Zn) plays role in wound healing [3,4]. Suboptimal intake of some elements can lead to disfunction of many essential bioactive processes. Supraoptimal concentration of several metals e.g., Fe, Zn, Cu, Al or Mn have toxic actions on nerve cells and neurobehavioral functioning, which can be expressed either as developmental effects or as an increased risk of neurodegenerative diseases in old age [5].

Ginkgo (*Ginkgo biloba* L.) preparations are mainly used for the treatment of peripheral vascular disease or cerebral insufficiency, especially by elders [6, 7, 8]. Accordingly, using whole ginkgo leaves as drug for tea preparations or tinctures, the element content and the dissolution rate of mineral elements in aqueous and alcoholic extracts may have a determining influence on bioactive effect in the body [9, 10].

Ginkgo is a dioecious plant, for cultivation of leaf drug only male trees are used. The aim of this chapter was to determine the element content of *Ginkgo biloba* leaves collected in different sites in Hungary, and to compare the element composition of male and female trees.

Experimental

Ginkgo (*Ginkgo biloba* L.) leaves were collected monthly during the vegetation period from May to September in three years following each other, from either male and female trees. Three collection places were chosen in Hungary (Europe), where the male and female trees are similar old and are planted near to each other: in the botanical garden of the Corvinus University of Budapest (Site 1), in that of the Eötvös Loránd University (Budapest) (Site 2) and in the main square of city Szeged (Site 3). Sample collection was carried out triplicated. Directly after collection leaves were dried at 30°C and than pulverized.

For analysis of element composition, 0.2 g of samples (duplicated) was taken. The following elements were determined by inductively coupled plasma optical emission spectrometry (ICP -OES): Al, As, B, Ba, Ca, Cd, Co, Cr, Cu, Fe, K, Li, Mg, Mn, Mo, Na, Ni, P, Pb, Se, Si, Sr, Ti, V, Zn. The concentration of the elements is given as mg kg^{-1} dry matter.

Conclusion

The concentration of As, Cd, Co, Cr, Ga, Hg, Mo, Pb and Se was under the detection limit in each samples. The absence of heavy metals could be an indication that the investigated samples are free of toxic metals.

Element content was determined in the dry matter of collected ginkgo leaves, concentrations of the main elements are given in Table 1. As there were no significant differences among the investigated years in the element content of ginkgo leaves, average values of the three years are summarized.

Significant differences were found among the collection places in the element concentrations of Al, Ca, Fe, K, Mg, Mn, Na, P, Sr, Zn. Uptake of some elements was depending on the chemical composition of soil and air at the observed places [1, 11].

Accumulation dynamism of the elements in the ginkgo leaves was quite diverse during the vegetation period. Accumulation of the most elements has risen continuously, however the concentrations of Al, B and Ca increased rapidly at the end of summer. During the vegetation period, concentration of Al, Ca, Mg increased by 1,5-2-fold, since content of B and Ba was more than 3-fold higher at the end of the investigation period.

The concentrations of calcium and magnesium showed a high variety, their level was high in ginkgo leaves compared to other species used for herbal tea [9]. This result is in agreement with the findings of other authors [10, 12, 13]. Ca content was measured in a wide range from 18320 to 40787 mg kg^{-1}, while Mg content ranged from 2503 to 6926 mg kg^{-1}, according to the collection place and the sex of the tree (Table 1). Calcium and magnesium levels changed seasonally, calcium and magnesium were lowest in the spring and rose about 1,5-2-fold to a maximum in the early autumn. This trend was pronounced in every collection place, and in case of both male and female trees. Calcium level was influenced some degree by the environmental conditions, however, for magnesium similar degree of accumulation was found at the different sites.

Potassium and phosphorus concentrations increased constantly during the vegetation period, potassium content varied in a wide range (10634 to 34102 mg kg^{-1}), and phosphorus content ranged from 2017 to 4128 mg kg^{-1}. Differences in the K and P concentrations could be detected at the different collection places, this result refers to the different avaibility of these elements in the soil [10,12]. Potassium level is very high in ginkgo leaves, what is especially favorable for human uses.

Table 1. Concentration of macro elements in ginkgo leaves from different sites (mg kg-1)*

	Male trees					Female trees				
	May	**June**	**July**	**August**	**September**	**May**	**June**	**July**	**August**	**September**
Site 1										
Ca	24746±135	28760±178	32702±143	38763±140	40225±166	19459±147	20480±169	26170±102	29622±124	30721±167
Mg	3427±25	3604±13	3631±29	4544±34	6369±38	3554±39	4685±15	5648±27	6356±23	6926±38
K	10634±177	10245±104	10610±99	10878±127	11062±164	12930±134	13530±119	18930±154	20588±143	21144±164
Na	121.7±5.1	141.3±6.7	165.4±6.3	152.4±7.7	154.4±4.1	130.1±4.2	132.6±7.1	137.7±5.2	147.5±5.9	142.8±6.5
P	2302±17	2318±32	2737±25	2753±12	2822±38	3018±41	3360±29	3212±34	3231±55	3265±23
Site 2										
Ca	20471±157	29482±120	33526±94	3959±116	40787±93	20078±94	25432±116	29840±127	35218±93	36540±104
Mg	3425±22	3480±18	4682±39	6543±51	6387±43	3655±29	3869±37	4770±22	6847±68	6784±79
K	19024±147	20114±139	21356±104	23359±98	25894±118	24354±112	26150±134	28397±106	34071±122	34102±136
Na	199.2±3.9	203.1±5.3	210.4±5.9	218.0±7.6	210.6±10.3	211.9±10.1	210.8±5.4	214.6±6.2	217.8±4.3	219.8±8.2
P	2017±36	2165±14	2473±23	2043±40	2170±28	2685±31	2867±17	2812±23	3016±26	3029±37
Site 3										
Ca	20520±123	21567±73	24841±78	36880±102	37045±88	18320±82	19452±73	24384±85	29847±115	32075±124
Mg	3246±40	3425±24	3510±36	3675±46	3752±104	2503±55	2536±29	2645±34	2983±61	3428±67
K	18240±113	19465±119	20475±100	23701±142	24086±122	21362±104	22347±91	23458±83	25363±92	26448±94
Na	88.9±4,3	89.4±2.8	90.2±5.7	93.9±8.6	94.5±12.7	87.3±4.5	89.2±4.6	95.8±7.2	99.4±5.3	103.5±7.8
P	3680±22	3696±19	3705±37	3789±35	3805±27	3956±18	3988±23	4017±43	4035±38	4128±42

*Values are means of three-years data ± standard deviation.

The concentration of sodium varied from 87.28 to 219.4 mg kg^{-1}. Na content was slightly depending on the collection place, while it was not influenced by the vegetation period. High sodium content in the body has been associated with high blood pressure [14], but this may not be possible in a situation of higher potassium content [15]. The ratio of sodium to potassium (Na/K) in the body is of great concern for prevention of high blood pressure, Na/K ratio less than 1 is recommended [16]. In all leaf samples a very low Na/K ratio was found.

Aluminum concentration was detected in the spring in a range from 30.52 to 52.13 mg kg^{-1}, and from 50.96 to 91.89 mg kg^{-1} in the autumn. Malik and co-workers [10] measured 5-10-fold higher Al content than these values. Aluminum level in the plants is influenced by the soluble aluminum concentration in the soil, hence great differences can be found in the Al content in plant samples from different sites.

Seasonal element accumulation, i.e., an increase of the element concentrations during the vegetation period was detected also for boron and barium at all investigated sites. Boron content changed from 41.32-48.43 mg kg^{-1} in May to 105.40-167.80 mg kg^{-1} in September, and concentration of barium increased from 10.38-19.52 mg kg^{-1} in the spring to 41.31-73.56 mg kg^{-1} in the autumn. Boron level of ginkgo leaves was higher than that of other herbal plants, however, barium concentration was similar to them [9].

Concentration of copper was detected on a steady low level (5.47-7.56 mg kg^{-1}), manganese content was measured from 13.58 mg kg^{-1} to 32.69 mg kg^{-1}. Our findings correspond to those reported by other authors [10, 17]. Copper and manganese level of ginkgo leaves seems to be much lower than that of other herbal drugs and of urban tree leaves [9, 18, 19].

The iron content varied in ginkgo leaves from 45.36 to 59.27 mg kg^{-1}, from 81.90 to 117.34 mg kg^{-1}, and from 79.31 to 177.50 mg kg^{-1} at the Site 1, 2, 3, respectively. Different values of other authors [10,12,17] lead to the supposition that also iron content of ginkgo leaves is highly influenced by the growing site.

Strontium level changed from 89.32 to 180.40 mg kg^{-1}, differences among the studied sites were significant. Content of zinc was measured between 10.04-18.32 mg kg^{-1}, this level is a little higher than detected by other authors [10, 17].

Element compositions of male and female trees were compared in every sampling time, eliminating the effect of phenophases. Significant differences were obtained between the sexes for some elements. Concentrations of Al, Fe, K, Na, P, Zn were higher in leaves of female trees than those in leaves of male trees. However, concentration of Ca seemed to be higher in the leaves of male ginkgo trees.

REFERENCES

[1] Soetan, KO; Olaiya, CO; Oyewole, OE. *Afr. J. Food Sci.*, 4, 200-222 (2010).

[2] Sposito, G. *The environmental chemistry of aluminum*, CRC Press Inc., Boca Raton, FL (1996).

[3] WHO, *Trace elements in human nutrition and health*, Word Health Organization (1996).

[4] Pors Nielsen, S. *Bone*, 3, 583-588 (2004).

[5] Halliwell, B. *J. Neurochem.*, 5, 1609-1623 (1992).

[6] Sasaki, K; Wada, K; Haga, M. *Stud. Nat. Prod. Chem.*, 28, 165-198 (2003).

[7] Chan, PC; Xia, Q; Fu, PP. *J. Environ. Sci Health C*, 25, 211-244 (2007).

[8] Van Beek, TA; Montoro, P. *J. Chromatogr. A*, 1216, 2002-2032 (2009).

[9] Razic, S; Onkia, A; Dogo, S; Slavkovic, L; Popovic, A. *Talanta*, 67, 233-239 (2005).

[10] Malik, J; Frankova, A; Drabek, O; Szakova, J; Ash, C; Kokoska, L. *Food Chem.*, 139, 728-734 (2013).

[11] Azimi, S; Ludwig, A; Thévenot, DR; Colin, JL. *Sci. Total Environ.*, 308, 247-254 (2003).

[12] Yao, X; Zhou, GS; Tang, YP; Shang, EX; Su, SL; Qian, DW; Duan, JA. *Guang Pu Xue Yu Guang Pu Fen Xi*, 3, 808-812 (2013).

[13] Jiang, B; Jiang, GB; Liu, CJ; Ma, K; Jin, LM; Gao, JM. *Guang Pu Xue Yu Guang Pu Fen Xi*, 3, 812-815 (2010).

[14] Dahl, LK; Leitl, G; Heine, M. *J. Exp. Med.*, 2, 318-330 (1972).

[15] Adrogué, HJ; Madias, NE. *N. Engl. J. Med.*, 356, 1966-1978 (2007).

[16] Food and Nutrition Board, Institute of medicine. *Dietary reference Intake for Energy, Carbohydrate, Fibre, Fat, Fatty Acids, Cholesterol, protein and Amino acid (micro-nutrients)*, National Academies Press, Washington, DC (2005).

[17] Wang, XY; Chen, J. *J. Changshu Inst. Technol.*, 4, 2008.

[18] Tomasevic, M; Anicic, M. *Facta Universitatis, Series Physics, Chemistry and Technology*, 8, 1-13 (2010)

[19] Khan, SA; Khan, L; Hussain, I; Marwat, KB; Akhtar, N. *Pak. J. Weed Sci. Res.*, 1-2, 101-110 (2008).

In: *Ginkgo biloba*
Editor: Emmett Fisher
ISBN: 978-1-63484-460-4

Chapter 3

DIFFERENT GINKGO PREPARATIONS

Noémi Koczka[*]
Institute of Horticulture, Szent István University, Gödöllő, Hungary

ABSTRACT

Ginkgo biloba L. has been used as a traditional Chinese herbal medicine for thousands of years. Several studies have already appeared on the ginkgo tree and its chemical constituents. As a result of the intense pharmacological and clinical research, phytopharmaceuticals based on partially purified ginkgo leaf extracts are now among the most sold drugs all over the world. Although many different components contribute to the overall pharmacological effect of ginkgo extracts, ginkgolides are considered to be responsible for a significant part of the beneficial effects. Standardized ginkgo leaf extracts (EGb), containing 24% flavonol glycosides and 6% terpene trilactones, are primarily prescribed for problems associated with a poor central and peripheral blood circulation like dementia, vertigo and tinnitus. Ginkgo is reported to be efficient for the treatment of Alzheimer's disease and cardiovascular disease.

Allergenic and toxic compounds, such as ginkgotoxin and ginkgolic acids were also found in ginkgo leaves. These constituents can be present in ginkgo leaf teas, while they are removed from products containing the standardized leaf extract. Recently dried ginkgo leaves are being commonly consumed on a daily basis for their stimulant properties.

Ginkgo products are sold mainly not as medicines but as dietary supplements due to the less strict regulation of herbal products

[*] E-mail address: koczka.noemi@mkk.szie.hu.

considering purity and potential. The pharmaceutical quality of the different ginkgo preparations is highly depending on the chemical composition of the ginkgo leaves, the extraction method and the dissolution rate. These all determine the bioavailability and efficacy of the different products.

The use of medicinal plants or natural products increased in the last decades all over the world. One of the most popular herbal plants is *Ginkgo biloba* L. because of its widespread healing effects. Ginkgo is often called a „living fossil", namely it is one of the oldest surviving plant species [1-4], and today it is one of the most widely used herbal remedies in the world.

Ginkgo has been used by the traditional medicine for thousands of years. It has been a sacred tree, a symbol of yin and yang, of longevity and vitality. The Chinese had begun their medicinal use of the tree with the seeds, but they soon added the leaves and even the roots to their list of valuable medicinal materials [5-6]. *Ginkgo biloba* has been thoroughly investigated for its constituents, and a whole array of compounds has been described. Modern research on the ginkgo began with chemical analysis of its active constituents during the 1920's [7-9], which was then followed by a continual line of research to the present.

Ginkgo seed is known to be a medication and a food. Seeds are widely used in Asian cooking, in stuffing, soups, desserts, meat and poultry dishes as well as many vegetarian dishes, and the roasted seed is a popular delicacy [5, 6, 9, 10, 11]. However, the ginkgo seed can be toxic if eaten in large quantities or over a long period, especially by children. Eating boiled ginkgo seeds is safer than eating them raw. The ginkgo seed contains a toxic material, MPN (4-methoxypyridoxine), called ginkgotoxin [12-13]. The concentration of ginkgotoxin in the albumen of ginkgo seeds increases during the vegetation period and reaches its maximum at the beginning of August, subsequently the ginkgotoxin content declines rapidly. Canned and boiled albumens contain only 1% of the ginkgotoxin level found in the raw seed [13]. However, the ginkgotoxin concentration in roasted seeds is not much lower than that of raw seeds since this compound is rather stable. In Asian countries people traditionally avoid eating too much of ginkgo seeds during a single meal [5, 6, 9].

The medicinal uses of ginkgo seed were mainly involved with treating lung diseases. One of the famous traditional Chinese formulas for treating asthmatic breathing, Ding Chuan Tang, has ginkgo seed as a major compound [1, 14]. The seed extract was documented to inhibit various bacteria, including

Mycobacterium, causative agent of tuberculosis. Ginkgo seed has also been used as an astringent to treat fluid discharges [8]. In China and Japan ginkgo preparations are used in the treatment of cough, bronchial asthma, irritable bladder, and even alcohol abuse [15-16].

Ginkgo leaves have long been used in China to treat a variety of ailments and conditions, including asthma, bronchitis, angina pectoris, fatigue, heart attack, and tinnitus [5, 9]. Leaves are also used to alleviate high cholesterol levels and high blood pressure [15]. Pharmacological investigations of ginkgo's active ingredients were reported soon after the main constituents were isolated. Clinical studies with ginkgo began during the 1960's in China with studies of the leaf extract in treatment of cardiovascular diseases. European research splayed after releasing of standardized ginkgo leaf extracts (EGb 761) from the German company Schwabe [17-18].

Flavonoids are major constituents of *Ginkgo biloba* leaves, including biflavones, flavones, flavonols and associated glycosides [19-20]. The main flavonoid aglycones found in ginkgo include quercetin, kaempferol, isorhamnetin, apigenin, and myricetin [21-24]. Terpenes are also important active ingredients of *G. biloba*: ginkgolides and bilobalide are its unique components [9, 29]. Ginkgolide A, B, C, J, K, L and M have a cage-like molecule structure with six five-membered rings and a tert-butyl group with a difference of the position and number of substituted hydroxyl groups on the spirononane framework [4, 30, 31]. Bilobalide is a sesquiterpene lactone, including a tert-butyl group and two hydroxyl groups in its chemical structure [32-33]. The leaves accumulate more terpenes than the roots and shoots [33]. In the leaves ginkgolide A is present at the highest concentration, followed by ginkgolide B and C, and a small amount of the other ginkgolides [4, 30, 33]. Ginkgolides and bilobalide are well soluble in polar and intermediately polar organic solvents like lower alcohols, tetrahydrofuran, acetone and ethyl acetate, while moderately soluble in diethyl ether and water and insoluble in non-polar solvents like chloroform, toluene and hexane [30]. The solubility in water increases significantly at higher temperatures and refluxing in water or water with a certain percentage of methanol has been a regularly used procedure [34].

The flavonoid glycosides and the terpene lactones (ginkgolide A, B, C and bilobalide) are considered as the relevant constituents for pharmacological effects of the ginkgo leaves [2]. Other constituents include proanthocyanadins, glucose, rhamnose, organic acids, and alkylphenols. Alkylphenols (e.g., ginkgolic acids, ginkgol, bilobol) possess contact allergenic reactions, cytotoxic, mutagenic and slight neurotoxic properties. Presence of these

compounds is considered undesirable, so they are removed from the extracts [2, 12, 35].

Standardized dry extract of *G. biloba* leaves (EGb 761) is refined and quantified to 22-27% ginkgoflavonglycosides content, represented by quercetin, kaempherol and isorhamnetin; 5-7% terpene lactones, represented by ginkgolides A, B, C (2.8-3.4%) and bilobalide (2.6-3.2%). Ginkgolide B and bilobalide account for about 0.8% and 3% of the total extract, respectively. The content of ginkgolic acids in the extract is less than 5 ppm. For the extraction acetone (60%) is used [36-38]. To attain this level of active ingredients requires concentrating them by about 25 times their natural levels.

Dry extracts of ginkgo are available in the form of capsules and film-coated tablets. The clinically effective dosage of the standardized commercial extract is 120-240 mg/day. Treatment time of 4-6 weeks is considered a minimum duration to observe improvements, with 3-6 months as a standard course of treatment for existing symptomatic diseases (6 months is usually the maximum duration of a clinical trial; longer use may be necessary to maintain the desired effects). Although adverse effects have not been reported during the use of standard dosages of ginkgo extract yet, a few people reported mild gastric disturbance or headache. Nevertheless, no investigation examined the effect of the use of daily high dosages (up to 600 mg) for a longer period of several months [39-44].

Ginkgo is considered to be effective in the case of several disorders affecting two fundamental aspects of human physiology: improving blood flow to the brain and other tissues; and enhancing cellular metabolism. Most of the illnesses are associated with old age, when both blood flow and cellular metabolism deteriorate. Other disorders that can respond to ginkgo treatments as asthma, vertigo, tinnitus can also occur during the whole lifetime.

Several studies reported that ginkgo extract promoted vasodilation and improved blood flow through arteries, veins and capillaries [11, 45-49]. Some results suggest that ginkgo leaf extract can be useful in preventing and treating cardiovascular diseases [50-54]. Some compounds of ginkgo also showed antitumor activity [55-57].

Several human clinical trials reported positive effects of ginkgo leaf extracts on the improvement of cognition, reduction of memory loss, or improved blood flow which can be beneficial for dementia, Alzheimer's disease, vertigo, tinnitus, and other neuropsychiatric disorders [58-67].

Ginkgo is proved to be effective against Alzheimer's disease [68]. Alzheimer's disease is a form of dementia that progressively deteriorates the intellectual capacity of various domains of the brain, particularly with aging.

Ginkgo leaf extract is known to inhibit the formation of amyloid beta peptide (Aβ) from β-amyloid precursor protein (APP), a crucial process in the pathogenesis of Alzheimer's disease. Alternatively, the ginkgo leaf extract inhibits reactive oxygen species (ROS) accumulation induced by Aβ (particularly flavonol quercitin) and also reduces neuron apoptosis, where apoptosis is considered to be one of the main causes for neurodegenerative diseases [68-69], and thus help to relieve Alzheimer's disease [70].

Other studies demonstrated benefits of ginkgo leaf in treating Parkinson disease, depression and schizophrenia [71-77].

Ginkgo extract inhibits platelet aggregation and prolongs bleeding time, a ginkgolides are antagonists of platelet-activating factor (PAF). Besides causing platelet activation and aggregation, PAF produces proinflammatory effects (eg, increasing vascular permeability), it is an extremely potent ulcerogen in the stomach, and contracts smooth muscle, including bronchial muscle. Platelet-activating factor has a direct effect on neuronal function and long-term potentiation [78-83].

Flavonoids and terpenoids contribute to ginkgo's antioxidant and free radical scavenger effects [18]. Ginkgo has been found to reduce cell membrane lipid peroxidation, to protect brain neurons against oxidative stress induced by peroxidation. The ginkgo leaf extract can scavenge ROS such as hydroxyl radicals ($OH^{\cdot}$), peroxyl radical ($ROO^{\cdot}$), superoxide anion radical ($O^{2-\cdot}$), nitric oxide radical ($NO^{\cdot}$), hydrogen peroxide (H_2O_2), and ferryl ion species [50, 84]. Ginkgo can also enhance activities of antioxidant enzymes such as superoxide dismutase, glutathione peroxidase, catalase, and/or heme-oxygenase-1, thereby indirectly contributes as an antioxidant [84-90].

However, several studies showed no positive effect of the ginkgo leaf extract on blood flow, cognitive functions, or neurological problems [91-96]. Studies querying the effectiveness of ginkgo dry extracts were carried out mainly using tablets and capsules containing EGb 761. The quality of clinical trials before the millenary were variable, many of them used unsatisfactory methods or small numbers of treatments. Only a few investigations were found to be well-designed and conducted with randomized placebo control [39].

The widespread use of the ginkgo extract can also cause herb-drug interactions, altering drug efficiency or leading to undesired toxic effects of concurrent medications, especially for drugs with narrow therapeutic indices. Several studies showed that *G. biloba* extract and its constituents could influence the pharmacokinetics of coadministered drugs via altering the expression and activity of drug-metabolizing enzymes and transporters. Clinical trials identified herb-drug interactions potentiated by the concurrent

use of ginkgo extract, these studies are reviewed by Chen and co-workers [97-98].

A number of review articles suggested that ginkgo leaf extract could increase the risk of bleeding, since it can reduce blood thrombotic ability. Potentially serious adverse effects associated with ginkgo have also been reported [56, 99-103].

G. biloba is contraindicated in patients with a history of hypersensitivity to ginkgo preparations. Due to insufficient information, the use of ginkgo leaf extract in pregnancy and lactation is not recommended.

Quality is a key issue in the development of herbal medicinal products that have consistent safety and efficacy. Problems in the development of herbal remedies include the frequent lack of standardized products, a lack of toxicology, pharmacokinetic and pharmacodynamic data, as well as of dose-response and interaction studies. In addition, the placebo effect in trials with herbal remedies is often very high.

Ginkgo preparations containing dry leaf extract (tablets, capsules) are marketed in the United States as a dietary supplement. In Europe, ginkgo is available either as a medicinal product or as a dietary supplement. The main differences between the two categories are quality requirements and legal classification. In Europe most herbal products were marketed as medicinal products. This changed when directives of the Council of the European Economic Community were implemented, with the requirement of quality, efficacy, and safety data on medicinal products. Numerous medicinal products were not able to meet those requirements and were nevertheless marketed as „dietary supplements", which are considered food products by the law. This is currently the case also for different ginkgo products, which are widely marketed as a food supplement of questionable quality. Consumers are also be exposed to danger when buying ginkgo products, such as tablets, capsules, teas, and cosmetics via internet, as most of them are released without any permission and are often produced without the expected quality control.

The next important preparation form of ginkgo leaves is the tea. Ginkgo leaves are available in mono teas and in tea mixtures, which are often consumed on a daily basis. While the intake of such products is usually limited through the recommended daily dosage of a standardized medication, this limitation does not apply to ginkgo teas. In theory, toxic ingredients like ginkgolic acid should be removed before the herb is used in any preparations. In standardized ginkgo leaf extracts the concentrations of these compounds are limited. Conversely ginkgo teas may contain a higher level of ginkgolic acids.

Nowadays, oral liquids or injectable solutions of ginkgo leaves can be also purchased. Fresh plant extracts might be especially efficient for the treatment of neurodegenerative problems [104]. Alcoholic extraction results in a higher level of active ingredients, hence a more efficient product. One of the most important indications for the usage of ginkgo leaves is the antioxidative protection. Antioxidant capacity of ginkgo leaves are higher in alcoholic extracts than in aqueous extracts, therefore tinctures can be more effective than water solutions of the leaves [105-108].

The increasing interest in alternative medical practices has led to a number of controlled studies on herbal and homeopathic agents. Some publications on the homeopathic use of ginkgo and a few small proofs of evidence exist, which suggest the efficacy of ginkgo leaves also in homeopathic preparations [109-111]. Cosmetic products containing ginkgo leaf extract are also highly sought after. Ginkgo extract promotes the cellular regeneration and capillary blood flow of the skin, and has a great anti-aging effect through its antioxidant properties. Ginkgo can help to protect skin from redness and inflammation during exposure to UVA/UVB light as well. Ginkgo leaf also seems to be capable of increase skin moisture content and reduce inflammatory factors in skin [112-114].

The pharmaceutical quality of the different ginkgo preparations is highly influenced by the chemical composition of the ginkgo leaves and the extraction method. The concentration of active ingredients and the element composition of ginkgo leaves depend on the growing conditions and harvesting time [105-106, 115-119]. Leaves of male and female trees contain different amounts of active ingredients, macro and micro elements [108, 117]. The dissolution and bioavailability of the active components of the oral solid preparations of different ginkgo preparations are reported to differ markedly, due to their limited solubility [120-121]. Since the quantities of biologically active components in ginkgo products have a broad range, their pharmacological effects might be considerably variable [120-126].

These facts indicate that the pharmaceutical properties of different ginkgo products have a significant impact on the rate and extent of drug absorption, and very likely on the effectiveness of the product in the prevention or treatment of diseases.

References

[1] P. Del Tredici, *Arnoldia Boston*, 4, 2-11 (1992).

[2] T. A. van Beek, *J. Chromatogr. A*, 967, 21-55 (2002).

[3] V. Melzheimer and J. J. Lichius, in: T. A. van Beek (Ed.), *Ginkgo biloba*, Harwood, Amsterdam (2000).

[4] P. F. Michel and D. Hosford, in: P. Braquet (Ed.) *Ginkgolides - Chemistry, Biology, Pharmacology and Chemical Perspectives*. Vol. 1, Prous Science Publisher, Barcelona (1988).

[5] M. Schmid and H. Schmoll, *Ginkgo. Urbaum und Arzneipflanze*, Wissenschaftliche Verlaggesellschaft, Stuttgart (1994).

[6] P. Del Tredici, *Arnoldia Boston*, 2, 2-15 (1991).

[7] L. M. Perry, *Medicinal Plants of East and Southeast Asia: Attributed Properties and Uses*. M.I.T. Press, Cambridge (1984).

[8] H. Y. Hsu, Y. P. Chen and M. Hong, *The Chemical Constituents of Oriental Herbs*, Vol. 1, Oriental Healing Arts Institute, Long Beach, CA (1982).

[9] H. Huh and E. J. Staba, *J. Herbs Spices Med. Plants*, 1-2, 91-124 (1992).

[10] R. Bilia, *Fitoterapia*, 73, 276-279 (2002).

[11] D. J. McKenna, K. Jones and K. Hughes, *Altern. Ther. Health Med.*, 5, 70-90 (2001).

[12] E. Leistner and C. Drewke, *J. Nat. Prod.* 73, 86-92 (2010).

[13] A. Arenz, M. Klein, K. Fiehe, J. Gross, C. Drewke, T. Hemscheidt and E. Leistner, *Planta Med.*, 62, 548-551 (1996).

[14] K. Chan, M. L. Kuo, J. J. Shen, L. C. See, H. H. Chang and J. L. Huang, *Ped. Allergy Immunology*, 17, 316-322 (2006).

[15] O. Sticher, B. Meier and A. Hasler, In: T. A. van Beek (Ed.) *Ginkgo biloba*, Harwood Academic Publishers, Amsterdam (2000).

[16] W. Blaschek, S. Ebel and E. Hackenthal, Hagers Handbuch der Drogen und Arzneistoffe, Springer, Heidelberg (2002).

[17] H. Wagner, S. Bladt, U. Hartmann, A. Daily and W. Berkulin, Dtsch. Apoth. Ztg. 45, 2421-2429 (1989).

[18] H. Schilcher, *Zeitschrift Phytother.* 9, 119-127 (1988).

[19] M. J. Dubbe and I. Kanfer, *J. Pharm. Pharmaceut Sci.* 7, 303-309 (2004).

[20] T. A. van Beek and P. Montoro, *J. Chromatogr. A*, 1216, 2002-2032 (2009).

[21] A. Wettstein, *Phytomedicine,* 6, 393 (2000).

[22] T. A. van Beek, *Ginkgo biloba.* Harwood Academic Publishers, Amsterdam (2000).
[23] E. K. Perry, A. T. Pickering, W. W. Wang, P. J. Houghton and N. S. L. Peny, *Pharm. Pharmacol,* 51, 527 (1999).
[24] P. Curtis-Prior, D. Vere and P. Fray, *J. Pharm. Pharmacol.* 51, 535 (1999).
[25] C. Victoire, M. Haag-berruier, J. P. Lobstein-Guth Balz and R. Anton, *Planta Med.* 54, 245-247 (1989).
[26] M. Vanbaelen and R. Vanhaelen-Fastre, *Planta Med.* 55, 202-204 (1989).
[27] A. Hasler, G. A. Gross, B. Meier and O. Sticher, *Phytochem.* 31, 1391-1394 (1992).
[28] A. Hasler, O. Sticher and B. Meier, *J. Chromatogr. A*, 605, 41-48 (1992).
[29] S. Jaracz, S. Malik and K. Nakanishi, *Phytochem.* 65, 2897-2902 (2004).
[30] T. A. van Beek, Bioorg. Med. Chem. 13, 5001-5012 (2005).
[31] Y. Wang, L. S. Sheng and F. C. Lou, *Acta Pharmacol. Sin.* 36, 606 (2001).
[32] K. Nakanishi, K. Habaguchi, Y. Nakadaira, M. C. Woods, M. Maruyama, R. T. Major, M. Alauddin, A. R. Patcl, K. Weinges and W. Bahr, *Am. Chem. Soc.* 93, 3544 (1971).
[33] K. Sasaki, K. Wada and M. Haga, *Stud. Nat. Prod. Chem.* 28, 165-198 (2003).
[34] T. A. van Beek, H. A. Scheeren, T. Rantio, W. C. Melger and G. P. Lelyveld, *J. Chromatogr.* 543, 375-387 (1991).
[35] H. Jaggy and E. Koch, *Pharmazie,* 10, 735-738 (1997).
[36] J. O'Reilly, in: T. A. van Beek (Ed.), *Ginkgo biloba*, Harwood Academic Publishers, Amsterdam (2000).
[37] K. Ndjoko, J. L. Wolfender and K. Hostettmann, J. Chromatogr. B Biomed. Sci. Appl. 2, 249-255 (2000).
[38] J. V. Smith and Y. Luo, *Appl. Microbiol. Biotechnol.* 64, 465-472 (2004).
[39] J. Birks and J. Grimley Evans, *Cochrane Database Syst. Rev.* 4, CD003120 (2002).
[40] L. S. Schneider, S. T. DeKosky, M. R. Farlow, P. N. Tariot, R. Hoerr and M. Kieser, *Curr. Alzheimer Res.* 5, 541-551 (2005).
[41] H. Allain, P. Raoul, A. Lieury, F. LeCoz, J. M. Gandon and P. d'Arbigny, *Clinical Therapeutics* 3, 549-558 (1993).

[42] D. O. Kennedy, A. B. Scholey and K. A. Wesnes, *Psychopharmac.* 4, 416-423 (2000).
[43] U. Rigney, S. Kimber and I. Hindmarch, *Phytother. Res.* 13, 408-415 (1999).
[44] H. Herrschaft, A. Nacu, S. Likhachev, I. Sholomov, R. Hoerr and S. Schlaefke, *J. Psychiatric Res.* 6, 716-723 (2012).
[45] J. C. Winter, *Physiol. Behav.* 63, 425-433 (1998).
[47] F. Jung, C., H. Kiesewetter and E. Wenzel, *Arzneimittelforschung* 5, 589-593 (1990).Y. Wu, S. Li, W. Cui, X. Zu, F. Wang and J. Du, *Planta Med.* 7, 624-628 (2007).
[49] Y. Wu, S. Li, W. Cui, X. Zu, J. Du and F. Wang, *Phytomedicine* 3, 164-169 (2008).
[50] G. B. Mahady, *J. Cardiovasc. Nurs.* 16, 21-32 (2002).
[51] W. Zhou, H. Chai, P. H. Lin, A. B. Lumsden, Q. Yao and C. Chen, *Cardiovasc. Drug Rev.* 4, 309-319 (2004).
[52] M. Rodríguez, L. Ringstad, P. Schäfer, S. Just, H. W. Hofer, M. Malmsten and G. Siegel, *Atherosclerosis* 2, 438-444 (2007).
[53] D. Gardner, M. Messina, L. D. Lawson and J. W. Farquhar, *Curr. Atherosclerosis Rep.* 6, 468-475 (2003).
[54] H. Satoh and S. Nishida, *Clin. Chim. Acta* 1-2, 13-22 (2004).
[55] H. Itokawa, N. Totsuka, K. Nakahara, M. Maezuru, K. Takeya, M. Kondo, M. Inamatsu and H. Morita, *Chem. Pharm. Bull.* 6, 1619-1621 (1989).
[56] P. C. Chan, X. Qingsu and P. P. Fu, *J. Environ. Sci. Health C,* 25, 211-244 (2007).
[57] J. W. Kang, J. H. Kim, K. Song, S. H. Kim, J. H. Yoon and K. S. Kim, *Phytother. Res.* 1, 77-82 (2010).
[58] W. Hopfenmuller, *Arzneimittelforschung* 44, 1005-1013 (1994).
[59] E. Hartley, L. Heinze, S. Elsabagh and S. E. File, *Pharmacol. Biochem. Behav.* 75, 711-720 (2003).
[60] W. Issing, P. Klein and M. Weiser, *J. Altern. Complement. Med.* 11, 155-160 (2005).
[61] R. Donfrancesco and L. Ferrante, *Phytomedicine* 14, 367-370 (2007).
[62] O. Napryeyenko and I. Borzenko, *Arzneimittelforschung* 57, 4-11 (2007).
[63] A. Scripnikov, A. Khomenko and O. Napryeyenko, *Wien. Med. Wochenschr.* 157, 295-300 (2007).
[64] C. Ramassamy, F. Longpre and Y. Christen, *Curr. Alzheimer Res.* 4, 253-262 (2007).

[65] A. Cieza, P. Maier and E. Pöppel, *Fortschritte der Medizin* 1, 5-10 (2003).

[66] H. Woelk, K. H. Arnoldt, M. Kieser and R. Hoerr, *J. Psychiatric Res.* 6,472-480 (2007).

[67] S. T. DeKosky, J. D. Williamson, A. L. Fitzpatrick, R. A. Kronmal, D. G. Ives, J. A. Saxton, O. L. Lopez, G. Burke, M. C. Carlson, L. P. Fried, L. H. Kuller, J. A. Robbins, R. P. Tracy, N. F. Woolard, L. Dunn, B. E. Snitz, R.. L. Nahin and C. D. Furberg, *JAMA* 19, 2253-2262 (2008).

[68] B. Ahlemeyer and J. Krieglstein, *Pharmacopsychiatry* 1, 8-14 (2003).

[69] U. Ergun, E. Yurtcu and M. A. Ergun, *Cell. Biol. Int.* 29, 717-720 (2005).

[70] Z. X. Yao, Z. Han, K. Drieu and V. Papadopoulos, *J. Nutr. Biochem.* 15, 749-756 (2004).

[71] U. Hemmeter, B. Annen, R., U. Brüderlin, M. Hatzinger, U. Rose and E. Holsboer-Trachsler, *Pharmacopsychiatry* 2, 50-59 (2001).

[72] S. F. Yang, Q. Wu, A. S. Sun, X. N. Huang and J. S. Shi, *Acta Pharmac. Sin.* 12, 1089-1093 (2001).

[73] M. S. Kim, J. I. Lee, W. Y. Lee and S. E. Kim, *Phytother. Res.* 8, 663-666 (2004).

[74] R. Bridi, F. P. Crossetti, V. M. Steffen, and A. T. Henriques, *Phytother. Res.* 5, 449-451 (2001).

[75] K. Tanaka, R. F. S. Galduróz, L. T. B. Gobbi, and J. C. F. Galduróz, *Curr. Neuropharmacol.* 4, 430-435 (2013).

[76] X. Y. Zhang, D. F. Zhou, P. Y. Zhang, G. Y. Wu, J. M. Su and L. Y. Cao, *J. Clin. Psychiatry* 11, 878-883 (2001).

[77] J. Rathbone, L. Zhang, M. Zhang, J. Xia, X. Liu, Y. Yang and C. E. Adams, *Br. J. Psychiatry* 5, 379-384 (2007).

[78] P. F. Smith, K. Maclenman and C. L. Darlington, *J. Ethnopharmacol.* 50, 131-139 (1996).

[79] E. Koch, *Phytomed.* 1-2, 10-16 (2005).

[80] H. J. Cho and K. S. Nam, *Biochem. Molec. Biol.* 5, 678-683 (2007).

[81] Y. Deng, W. Fang, Y. Li, J. Cen, F. Fang, P. Lu, S. Gong and L. Mao, *Eur. J. Pharmacol.* 1-3, 43-47 (2009).

[82] V. Lamant, G. Mauco, P. Braquet, H. Chap and L. Douste-Blazy, *Biochem. Pharmacol.* 17, 2749-2752 (1987).

[83] S. Akiba, T. Kawauchi, T. Oka, T. Hashizumen and T. Sato, *Biochem. Molec. Biol. Int.* 6, 1243-1248 (1998).

[84] V. DeFeudis, V. Papadopoulos and K. Drieu, *Fundam. Clin. Pharmacol.* 17, 405-417 (2003).

[85] W. Song, H. U. Guan, X. Z. Zhu, Z. L. Chen, M. L. Yin and X. F. Cheng, *Acta Pharmacol. Sin.* 21, 415-420 (2000).
[86] I. Maitra, L. Marcocci, M. T. Droy-Lefaix and L. Packer, *Biochem. Pharmacol.* 11, 1649-1655 (1995).
[87] S. Pietri, J. R. Séguin, P. d'Arbigny, K. Drieu and M. Culcasi, *Cardiovasc. Drugs Ther.* 2, 121-131 (1997).
[88] N. Haramaki, S. Aggarwal, T. Kawabata, M. T. T. Droy-Lefaix and L. Packer, *Free Rad. Biol. Med.* 6, 789-794 (1994).
[89] A. Louajri, S. Harraga, V. Godot, G. Toubin, J. P. Kantelip and P. Magnin, *Biol. Pharmaceut. Bull.* 6, 710-712 (2001).
[90] M. Ellnain-Wojtaszek, Z. Kruczynski and J. Kasprzak, *Fitoterapia* 74, 1-6 (2003).
[91] M. C. van Dongen, E. van Rossum, A.G. Kessels, H. J. Sielhorst and P. G. Knipschild *J. Am. Geriatr. Soc.* 48,:1183-1194 (2000).
[92] M. van Dongen, E. van Rossum, A. Kessels, H. Sielhorst and P. Knipschild, *J. Clin. Epidemiol.* 56, 367-376 (2003).
[93] J. Carlson, J. W. Farquhar, E. DiNucci, L. Ausserer, J. Zehnder, D. Miller, K. Berra, L. Hagerty and W. L. Haskell, *J. Am. Diet. Assoc.* 107, 422-432 (2007).
[94] Lovera, B. Bagert, K. Smoot, C. D. Morris, F. R, K. Bogardus, K. Wild, B. Oken, R. Whitham R. and D. Bourdette, *Mult. Scler.* 13, 376-385 (2007).
[95] O. Lingærde, A. R. Fφreland and A. Magnusson, *Acta Psych. Scan.* 100, 62-66 (1999).
[96] E. Ernst, *Adv. Psychiatric Treatment*, 4, 312-316 (2007).
[97] X. W. Chen, E. S. Serag, K. B. Sneed, J. Liang, H. Chew and S. Y. Pan, *Curr. Med. Chem.* 31, 4836-4850 (2011).
[98] X. W. Chen, K. B. Sneed, S. Y. Pan, C. Cao, J. R. Kanwar and H. Chew, *Curr. Drug Metab.* 5, 640-651 (2012).
[99] S. Bent, H. Goldberg, A. Padula and A. L. Avins, *J. Gen. Intern. Med.* 7, 657-661 (2005).
[100] J. Rowin and S. L. Lewis, *Neurology* 6, 1775-1776 (1996).
[101] M. Rosenblatt and J. Mindel, *N. Engl. J. Med.* 15, 1108 (1997).
[102] D. Hauser, T. Gayowski and N. Singh, *Transpl Int.* 7, 377-379 (2002).
[103] S. Kohler, P. Funk and M. Kieser, *Blood Coagul. Fibrinolysis* 4, 303-309 (2004).
[104] A. Suter, W. Niemer and R. Klopp, *Adv. Ther.* 12, 1078-1088 (2011).
[105] M. Goh, P J. Barlow and C. S. Yong, *Food Chem.* 82, 275-282 (2003).

[106] J. Kobus, E. Flaczyk, A. Siger, M. Nogala-Kalucka, J. Korczak and R. B. Pegg, *Eur. J. Lipid Sci. Techn.* 111, 1150-1160 (2009).
[107] J. Ronowicz, B. Kupcewicz and E. Budzisz, *Centr. Eur. J. Biol.* 8, 374-385 (2013).
[108] N. Koczka, Z. Móczár, É. Stefanovits-Bányai and A. Ombódi, *Acta Pharm.* 2, 207-213 (2015).
[109] W. Issing, P. Klein and M. Weiser, *J. Altern. Complem. Med.* 1, 155-160 (2005).
[110] B. Schneider, P. Klein and M. Weiser, *Arzneimittelforschung* 1, 23-29 (2005).
[111] M. Nilsson, G. Trehn and K. Asplund, *J. Intern. Med.* 250, 225-233 (2001).
[112] S. dal Belo, L. Gaspar and P. Maia Campos, *Phytother. Res.* 12, 1854-1860 (2011).
[113] S. E. dal Belo, L. R. Gaspar P. M. Maia Campos and J. P. Marty, *Skin Pharmacol. Physiol.* 6, 299-304 (2009).
[114] M. D. Gianeti and P. M. Maia Campos, *Molecules* 19, 18268-18282 (2014).
[115] J. Malik, A. Frankova, O. Drabek, J. Szakova, C. Ash and L. Kokoska, *Food Chem.* 139, 728-734 (2013).
[116] N. Koczka, A. Ombódi, Z. Móczár and E. Stefanovits-Bányai, *Acta Alim.* in press.
[117] É. Stefanovits-Bányai, K. Szentmihályi, A. Hegedűs, N. Koczka, L. Váli, G. Taba and A. Blázovics, *Life Sci.* 78, 1049-1056 (2006).
[118] B. D. Xie and H. T. Wang, *J. Nanjing Forestry Uni.* 2, 51-54 (2006).
[119] L. Li, S. L. Tian and F. Zheng, *J. Anhui Agri. Sci.* 11, 2370-2414 (2006).
[120] S. Kressmann, W. E. Muller and H. H. Blume, *J. Pharm. Pharmacol.* 54, 661-669 (2002).
[121] S. Kressmann, A. Biber, M. Wonnemann and B. Schug, *J. Pharm. Pharmacol.* 54, 1507-1514 (2002).
[122] K. Woelkart, E. Feizlmayr, P. Dittrich, E. Beubler, F. Pinl, A. Suter and R. Bauer, *Phytother. Res.* 24, 445-450 (2010).
[123] F. Deng and S. W. Zito, *J. Chromatogr. A.* 1, 121-123 (2003).
[124] D. Mantle, R. M. Wilkins and M. Asim Gok, *J. Altern. Complem. Med.* 5, 625-629 (2003).
[125] Y. Kakigi, T. Hakamatsuka, T. Icho, Y. Goda and N. Mochizuki, *Biosci. Biotechnol. Biochem.* 4, 777-779 (2011),
[126] C. Guidetti, S. Paracchini, S. Lucchini, M. Cambieri and F. Marzatico, *J. Pharm. Pharmacol.* 53, 387-392 (2001).

In: *Ginkgo biloba*
Editor: Emmett Fisher
ISBN: 978-1-63484-460-4

Chapter 4

GINKGO BILOBA (GINKGOACEAE) AS A SOURCE OF PHENOLIC COMPOUNDS WITH ANTIOXIDANT ACTIVITY

Milan S. Stanković *
Department of Biology and Ecology, Faculty of Science,
University of Kragujevac, Republic of Serbia

ABSTRACT

This chapter presents the results of a screening of different crude extracts, infusions and standardized extract from the *Ginkgo biloba* (Ginkgoaceae) leaves for total phenolic content, concentration of flavonoids and *in vitro* antioxidant activity. Main reason for this study is the determination of these parameters and their variability among the plant extracts obtained by different solvents and water infusions prepared using different methods respectively, as well as standardized extract. Results for total phenolic content determined using Folin-Ciocalteu reagent and expressed in term of gallic acid equivalent, GAE (mg of GA/g of extract) ranged from 27.47 ± 0.29 to 141.60 ± 0.36 mg of GA/g of extract and plant material. The concentrations of flavonoids determined using spectrophotometric method with aluminum chloride and expressed in terms of rutin equivalent, RuE (mg of Ru/g of extract) ranged from 14.46 ± 0.23 to 231.15 ± 0.17 mg of Ru/g of extract and plant material. Obtained results for antioxidant activity of *Ginkgo biloba*

* E-mail address: mstankovic@kg.ac.rs.

extract and infusions ranged from 1408. 96 ± 2.01 to 49.75 ± 0.90 μg/ml. Results for antioxidant activity and the amount of total phenolics content and flavonoids varied according to the type of solvent used, as well as the method of preparation of the extracts and infusion. In addition, it can be concluded that there is a relation between the quantity of phenolics, flavonoids and antioxidant activity. Great variability of the studied parameters was observed comparing the effectiveness of the used solvents. The ethanolic extracts, infusion, infusion prepared with boiled water, as well as ethanolic solution of standardized *Ginkgo biloba* extract contain the greatest concentrations of phenolics compounds, especially flavonoids and showed high antioxidant activity. According to our research, leaves from *Ginkgo biloba* are rich sources of phenolic compounds with strong antioxidant activity.

Keywords: *Ginkgo biloba* (Ginkgoaceae), phenolics, antioxidant activity

INTRODUCTION

In addition to the primary, the secondary metabolism takes place in plant cells. Its products, secondary metabolites, are not essential for the plant, but instead, they represent, in most of the cases, the result of the plant adaptations to the environmental conditions [1, 2]. According to the chemical composition criterion, secondary metabolites can be divided into two categories [3, 4]. The molecules which do not contain the nitrogen atoms, such as phenolic compounds and terpenoids, belong to the first category, whereas the second group encompasses the metabolites whose molecules contain nitrogen atoms i.e., alkaloids. Biosynthetic pathways of secondary metabolites are extremely complex and a multitude of products and enzyme controlled reactions are part of them [5, 6]. Apart from having an important biological role in ecosystemic plant interactions, secondary metabolites isolated from the plant organism exhibit biological activity in both *in vitro* and *in vivo* conditions [7, 8]. The biological activity of plant secondary metabolites is derived from their capability to react with molecules and cell and subcell structures thus influencing, either positively or negatively, a great number of metabolic processes. Through the stimulatory and inhibitive mechanism, secondary metabolites display antioxidant, antimicrobial, antiproliferative, apoptotic, anti-inflammatory, antihypertensive, neuroprotective and many other activities [9-13].

Free radicals are the molecules, ions or atoms of carbon, oxygen, nitrogen, or sulfur with unpaired electrons which cause them to be quite reactive. Free radicals arise during a great number of metabolic processes in an organism and under the influence of negative effects of physical and chemical factors in the environment. The following reactions bring about the formation of free radicals: homolytic bond fission, oxidation-reduction reaction, thermolysis, photolysis, radiolysis, transfer of an electrone to an organic molecule, the activity of ozone, nitrogen(IV)-oxide and singlet oxygene and various enzyme processes in an organism [14]. The free radical and non-radical forms which most often arise and are characterized by the greatest degree of reactivity are reactive oxygen species such as superoxide anion radicals, hydroxyl radicals, hydroperoxyl radical, peroxyl radicals, singlet oxygen, carbon dioxide radicals and carbon monoxide radicals [15]. The mechanism of negative effect of free radical species in the biological systems is based on their reaction with biomolecules such as nucleic acids, proteins, lipids and enzymes which simultaneously causes their damage and disables their primary biological roles in cells. The increased production of free radicals in human organism intensifies ageing process of an organism and causes a great number of pathological states such as neurodegenerative changes, cancer genesis, cardiovascular disorders as well as inflammatory processes [16, 17]. The damaging effects of free radicals are inhibited by antioxidant substances. Antioxidants either partially disable or completely prevent the process of oxidation of a substrate. The mechanism of the antioxidant activity is derived from several capabilities antioxidants may have: the ability to donate electrons or hydrogen atoms i.e., scavenging ability, the ability to chelate ions of metals (Fe^{2+}, Cu^{2+}, Zn^{2+}, Mg^{2+}) thus decreasing their redox potential, and the ability to destroy hydroperoxides of lipid molecules from which non-radical species develop. To this end, there is a great number of natural (enzymatic and non-enzymatic) and synthetic antioxidant substances. Enzymatic antioxidants are catalase, peroxidase, superoxide dismutase, glutathione reductase, glutathione peroxidase and glutathione S-transferase wheras α-tocopherol, β-carotene and phenolic compounds isolated from the plants belong to the category of non-enzymatic antioxidants [18, 19].

The usage of natural antioxidants does not cause adverse effects, whereas, in the case of synthetic antioxidants, genotoxic effect has been proven in addition to some other negative effects [20-22]. For this reason, there have been conducted numerous researches of the biological activity and chemical composition of natural products, and of medicinal plants, fruit and vegetable as potential sources of natural antioxidants in particular [23-26]. Antioxidant capability of phenolic compounds depends upon the number and the position of hydroxy groups. Apart from this, very important physical-chemical characteristics are bond dissociation energy and ionization potential. One of the main mechanisms of the antioxidant activity of phenolic compounds is the role in the transfer of hydrogen atom and in the transfer of an electrone to the molecule exposed to the oxidation process [6].

Ginkgo biloba (Ginkgoaceae) – ginkgo, ginko, silver fruit, silver peach or maidenhair tree is a plant known for several reasons, represents endemorelict or "living fossil" species and famous plant in folk and modern medicine. *Ginkgo biloba* is naturally widespread in the southwestern parts of the territory of China, because the distribution throughout the Northern Hemisphere was reduced during the last glacial period. After the intensive worldwide distribution, using different ways of cultivation, *Ginkgo biloba* is the object of numerous research studies in different fields of biological and pharmaceutical sciences, as well as widely used plant species as a source of bioactive substances with significant therapeutically effects. Conducted studies mostly deal with detailed identification of quantitative and qualitative composition of secondary metabolites, as well as *in vitro* and *in vivo* biological effects of the active secondary metabolites from different *Ginkgo biloba* plant organs [27].

Qualitative analysis of secondary metabolites of *Ginkgo biloba* was examined in a number of studies. In a detailed review [28] summarized secondary metabolites are: Terpenes (Monoterpenes: cymene, isopropyl-phenol, thymol, linalool oxide and ionone; Diterpenes: ginkgolide A, ginkgolide B, ginkgolide C, ginkgolide J, ginkgolide M, ginkgolide K and ginkgolide L; Sesquiterpenes: bilobalide, bilobanone, E- and Z- forms of 10-11-dihydroatlantone, E-10-11-dihydro-6-oxoatlantone, elemol and eudesmol; Steroids and phytosterols: β-sitosterol, stigmasterol, campesterol and dihydrobrassicasterol; Carotenoids: α-carotene, γ-carotene, lutein and zeaxanthin; Polyprenols: Ginkgo polyprenols and Ginkgo polyprenol acetates); Flavonoids (Glycosides of: kaempferol, quercetin, myricetin, apigenin, isorhamnetin and luteolin; Aglycones: kaempferol, quercetin, myricetin, apigenin, isorhamnetin, luteolin, tamarixetin, 4′-OMe apigenin, 3′-methylmyricetin, catechin, epicatechin, epigallocatechin and gallocatechin; Dimers: catechin–catechin, epicatechin–catechin, epigallocatechin–catechin and gallocatechin–catechin; Anthocyanidins: procyanidin and prodelphinidin; Biflavones: amentoflavone, 7-methoxyamentoflavone, bilobetin, 5′-methoxybilobetin, sequojaflavone, ginkgetin, isoginkgetin and sciadopitysin; Biflavone glucosides: ginkgetin and isoginkgetin); Alkyl phenols and alkyl phenolic acids (Cardanols: 3-tridecylphenol, 3-tetradecylphenol, 3-

pentadecylphenol, 3-heptadecylphenol and ginkgol; Cardols: 5-tridecylresorcinol, 5-tetradecylresorcinol, 5-pentadecylresorcinol, 5-heptadecylresorcinol and bilobolol; Anacardic acids: 6-tridecylsalicyclic acid, 6-tetradecylsalicylic acid, 6-pentadecylsalicylic acid, 6-[8-pentadecenyl] salicylic acid, 6-hexadecylsalicylic acid, 6-[9, 12-heptadecadienyl] salicylic acid and 6-[8-heptadecenyl] salicylic acid; Resorcylic acids: 6-[8-Pentadecenyl] resorcylic acid, 6-tridecylresorcylic acid and alkyl coumarin); Organic acids (ascorbic acid, D-glucaric acid, quinic acid, shikimic acid, 6-hydroxykynurenic acid, p-hydroxybenzoic acid, protocatechuic acid, vanillic acid, acetic acid, butyric acid, formic acid, hexanoic acid and valeric acid); carbohydrates, long chain hydrocarbons and lipids, inorganic salts or their complexes etc.

Among the many components identified in the aboveground parts of *Ginkgo biloba*, phenolic acids, flavonoid glycosides, terpene trilactones, ginkgolides and bilobalides exhibit a significant effect [29, 30]. Due to the content of the active components, *Ginkgo biloba* extract is standardized in the industrial production of drugs. As the main raw material for the extracts obtaining are leaves and seeds without sarcotesta (outer layer of the ovules).

Active substances from *Ginkgo biloba*, have an individual effect, or in most cases, synergistic effect. The mechanism of action of the *Ginkgo biloba* active compounds, was investigated in a number of studies, as well as by the application of different methodological approaches. Based on current results, the most important pathways of action are: effects on blood circulation such as vasoregulating activity of arteries, capillaries, veins (increased blood flow) and rheological effects (decreased viscosity, antagonistic to platelet activating factor receptors), metabolic changes, for example on neuron metabolism (increased tolerance for anoxia), beneficial influence on neurotransmitter disturbances and prevention of damage of membranes caused by free radicals [31].

Active ingredients of *Ginkgo biloba*, through these mechanisms of action, can eliminate the consequences of reduced circulation in peripheral parts of the body and in the brain, such as decreasing of concentration and memory capacity, dizziness, anxiety, headache, numbness and coldness of hands and feet, as well as tiredness and loss of will and energy. Because of these effects, pharmaceutical products with *Ginkgo biloba* extract, are used for the treatment of dementia and Alzheimer's disease, as a dietary supplement in the diet of healthy people, for memory improving, circulation, heart and brain infarction as well as for improving the antioxidant capacity of the organism [32-40].

Apart from numerous studies of neuroprotective activity and mechanisms of action, other levels of biological activities have been extensively investigated so far. During a number of previous *in vitro* and *in vivo* studies of *Ginkgo biloba* crude and standardized extracts of leaves, stem bark and seeds or essential oils, different aspects of the biological activity were evaluated, such as antimicrobial and antifungal potential and mechanisms of action [41-45], antioxidant activity [46-48], antimutagenic activity [49, 50], anticancer activity [51, 52], as well as hepatoprotective effect [47, 53].

This chapter deals with quantitative characteristics of phenolic compounds as the most important active substances and biological activity of *Ginkgo biloba* (Ginkgoaceae) plant extracts and infusions prepared from leaves. Selected methods are applied in the study of plant extracts obtained using different solvents, standardized extract prepared as industrial basic substance for drug production, as well as infusions prepared by different methods. The concentration of phenolics in the plant extracts and infusions of *Ginkgo biloba* leaves was measured by Folin-Ciocalteu reagent and expressed in terms of gallic acid equivalent, GAE (mg of GA/g of extract). The content of flavonoids in the examined plant extracts and infusions was determined using spectrophotometric method by aluminium chloride as reagent and expressed in terms of rutin equivalent, RUE (mg of RU/g of extract). After quantification of total phenolics as well as flavonoids separately, as the most important group of phenolic compounds, antioxidant activity of all samples of plant extracts and infusions from *Ginkgo biloba* leaves was determined using free radical (DPPH) assay. The results for antioxidant activity of analyzed *Ginkgo biloba* plant extracts and infusions were expressed in terms of IC_{50} (μg/ml) values.

MATERIALS AND METHODS

Plant Material

Ginkgo biloba leaves were collected from cultivated individuals (Kragujevac, Republic of Serbia). The voucher specimen was confirmed and deposited at the Herbarium of the Faculty of Sciences, University of Kragujevac. Plant material was air-dried in the dark, at ambient temperature. Air-dried material was milled in a grinder and stored in tightly sealed dark containers until the analysis.

Chemicals

Organic solvents and sodium hydrogen carbonate were purchased from „Zorka pharma" Šabac, Serbia. Gallic acid, rutin hydrate and 2,2-dyphenyl-1-picrylhydrazyl (DPPH) were obtained from Sigma Chemicals Co., St Louis, MO, USA. Folin-Ciocalteu phenol reagent and aluminium chloride hexahydrate ($AlCl_3$ x $6H_2O$) were purchased from Fluka Chemie AG, Buchs, Switzerland. All other solvents and chemicals were of analytical grade. A standardized extract of *Ginkgo biloba* was obtained from Pharmaceutical Company „Ivančić i Sinovi", Belgrade, Serbia (base for dietary products *Ginkgo biloba* extract, produced by Sichuan Xieli Pharmaceutical. Co. Ltd., Sichuan, China).

Statistical Analysis

All experimental measurements were carried out in triplicate and obtained results are expressed as average of three analyses ± standard deviation. Statistical analysis was done using a SPSS (Chicago, IL) statistical software package (SPSS for Windows, ver. 17, 2008). The concentrations wich neutralised 50% of free radicals (IC_{50} values) were calculated using software Origin 8 Pro (OriginLab Corp.).

Preparation of Plant Extracts and Infusions

For *Ginkgo biloba* leaves plant extracts (samples 1 – 7) preparation, 10 g of powder obtained from dry leaves was transferred into dark-coloured flasks, filled with 200 ml of solvent (water, methanol, ethanol, acetone, ethyl acetate, buthanol and petroleum ether) and stored at room temperature. After 24 h, infusions were filtered using Whatman No. 1 filter paper and residue was re-extracted with an equal volume of solvents. After 48 h, the process was repeated. Combined supernatants were evaporated to dryness under vacuum at 40°C using Rotary evaporator. For methanolic and ethanolic infusions (samples 8 and 9), 20 mg of powdered dry leaves was transferred into dark-coloured flasks, filled with 20 ml of solvent (methanol and ethanol, acetone, ethyl) and stored at room temperature. After 24 h, infusions were filtered using Whatman No. 1 filter paper. For water infusions preparation (samples 10 – 14), 20 mg of powdered dry leaves was transferred into dark-coloured flasks, filled with 20 ml of water of different temperatures (10°C for sample number 10, 50°C for samples number 11 and 13, as well as 100°C for samples number 12 and 14) and stored at room temperature. After 1 h for samples number 13 and 14, as well as 24 h for samples number 10, 11 and 12 infusions were filtered using Whatman No. 1 filter paper. Samples number 15 and 16 contain methanolic and ethanolic solution of standardized *Ginkgo biloba* extract in concentration of 1 mg/ml. The obtained extracts and infusions were kept in sterile sample tubes and stored in a refrigerator at 4°C.

Determination of Total Phenolic Content

The concentration of phenolics in the plant extracts was measured by using spectrophotometric method [54]. The methanol solution of the extract in concentration of 1 mg/ml was used in the analysis. The reaction mixture was prepared by mixing 0.5 ml of methanol solution of the extract, 2.5 ml of 10% Folin-Ciocalteu reagent dissolved in water and 2 ml of 7.5% $NaHCO_3$.

The blank was concomitantly prepared containing 0.5 ml of methanol, 2.5 ml of 10% Folin-Ciocalteu reagent dissolved in water and 2 ml of 7.5% of $NaHCO_3$. The samples were thereafter incubated at 45°C for 45 min. The absorbance was determined using spectrophotometer at λ_{max} = 765 nm.

The samples were prepared in triplicate for each analysis and the mean value of absorbance was obtained. The same procedure was repeated for the standard solution of gallic acid and the calibration line was construed. Based on the measured absorbance, the concentration of phenolics was read (mg/ml) from the calibration line; then the content of phenolics in the extracts was expressed in terms of gallic acid equivalent, GAE (mg of GA/g of extract).

Determination of Flavonoid Content

The content of flavonoids in the examined plant extracts was determined using spectrophotometric method [55]. The sample contained 1 ml of methanol solution of the extract in concentration of 1 mg/ml and 1 ml of 2% $AlCl_3$ solution dissolved in methanol. The samples were incubated for an hour at room temperature. The absorbance was determined using spectrophotometer at of λ_{max} = 415 nm.

The samples were prepared in triplicate for each analysis and the mean value of absorbance was obtained. The same procedure was repeated for the standard solution of rutin and the calibration line was construed. Based on the measured absorbance, the concentration of flavonoids was read (mg/ml) on the calibration line; then, the content of flavonoids in extracts was expressed in terms of rutin equivalent, RUE (mg of RU/g of extract).

Evaluation of DPPH Scavenging Activity

The ability of the plant extract and reference substance to scavenge DPPH (1,1-diphenyl-2-picrylhydrazyl radical) free radicals was assessed using the method described by Tekao et al. [56], adopted with suitable modifications from Kumarasamy et al. [57].

The stock solution of the plant extract was prepared in methanol to achieve the concentration of 1 mg/ml. Dilutions were made to obtain concentrations of 500, 250, 125, 62.5, 31.25, 15.62, 7.81, 3.90, 1.99, 0.97 μg/ml. Diluted solutions (1 ml each) were mixed with 1 ml of DPPH methanolic solution. After 30 min in darkness at room temperature, the absorbance was recorded at 517 nm.

The control samples contained all the reagents except the extract. The percentage inhibition was calculated using the equation: % inhibition = 100 x (A *of control* – A *of sample*)/A *of control*), whilst IC_{50} values were estimated from the % inhibition versus the concentration sigmoidal curve, using a non-linear regression analysis. In presented results, antioxidant efficiency of the extract increased with the decreasing of IC_{50} values. The data were presented as mean values ± standard deviation (n = 3).

Results and Discussion

The results of total phenolics content in *Ginkgo biloba* leaves, determined by measuring of the amount in the extracts and infusions using Folin-Ciocalteu's reagent, are presented in Table 1.

Obtained results for total phenolics content ranged from 27.47 ± 0.29 to 141.60 ± 0.36 mg of GA/g of extract and plant material. Total phenolics contents in the extracts obtained by different solvents and evaporation (1 – 7) ranged from 33.25 ± 0.19 mg of GA/g for petroleum ether extract to 82.00 ± 0.37 mg of GA/g for ethanolic extract. Apart from the ethanolic extract, high amount of total phenolics was measured in methanolic (71.02 ± 0.45 mg of GA/g) and buthanolic (63.27 ± 0.69 mg of GA/g) extract. The results indicate that solvent polarity has a great importance for the quantity of phenolics. The obtained results for the amount of total phenolic content in the organic solvent infusions (8 and 9) are 29.83 ± 0.07 mg of GA/g for methanolic and 36.73 ± 0.27 mg of GA/g for ethanolic infusion. Total phenolics content in the water infusions (10 – 14) obtained using different methods for preparation ranged from 27.47 ± 0.29 mg of GA/g to 34.74 ± 0.64 mg of GA/g. The results indicate that the temperature of the water and the time of preparation of water infusions, have an effect on the extraction process and the final amount of total phenolic compounds in the water solution.

Table 1. Total phenolic content in the Ginkgo biloba leaves extracts expressed in terms of gallic acid equivalent, GAE (mg of GA/g of extract or plant material)

Type of extract	mg of GA/g
1. Water extract	40.32 ± 0.53
2. Methanolic extract	71.02 ± 0.45
3. Ethanolic extract	82.00 ± 0.37
4. Acetone extract	58.36 ± 0.29
5. Ethyl acetate	46.72 ± 0.10
6. Buthanolic extract	63.27 ± 0.69
7. Petroleum ether	33.25 ± 0.19
8. Methanolic infusion	29.83 ± 0.07
9. Ethanolic infusion	36.73 ± 0.27
10. Water infusion (10°C, 24 h)	27.47 ± 0.29
11. Water infusion (50°C, 24 h)	33.22 ± 0.49
12. Water infusion (100°C, 24 h)	33.80 ± 0.51
13. Water infusion (50°C, 1 h)	29.22 ± 0.41
14. Water infusion (100°C, 1 h)	34.74 ± 0.64
15. Standardized extract dissolved in methanol	127.46 ± 0.30
16. Standardized extract dissolved in ethanol	141.60 ± 0.36

The results for the amount of total phenolics content in the standardized extract of *Ginkgo biloba* leaves (15 – 16) are 127.46 ± 0.30 mg of GA/g for methanolic and 141.60 ± 0.36 mg of GA/g for ethanolic solution. The differences in amounts between these extracts point to the importance of the type of the selected solvent. In this case ethanol contributes to better dissolution of phenolic compounds and their larger amount in the final sample.

The results of flavonoid content in *Ginkgo biloba* leaves, determined by measuring of the amount in the extracts and infusions using aluminium chloride as reagent, are presented in Table 2.

Table 2. Flavonoid content in the Ginkgo biloba leaves extracts expressed in terms of rutin equivalent, RuE (mg of Ru/g of extract or plant material)

Type of extract	mg of Ru/g
1. Water extract	27.59 ± 0.94
2. Methanolic extract	47.70 ± 0.18
3. Ethanolic extract	67.65 ± 0.29
4. Acetone extract	44.14 ± 0.15
5. Ethyl acetate	46.34 ± 0.19
6. Buthanolic extract	47.16 ± 0.44
7. Petroleum ether	15.94 ± 0.25
8. Methanolic infusion	26.66 ± 0.31
9. Ethanolic infusion	33.67 ± 0.28
10. Water infusion (10°C, 24 h)	14.46 ± 0.23
11. Water infusion (50°C, 24 h)	14.86 ± 0.36
12. Water infusion (100°C, 24 h)	17.92 ± 0.30
13. Water infusion (50°C, 1 h)	14.80 ± 0.39
14. Water infusion (100°C, 1 h)	15.36 ± 0.24
15. Standardized extract dissolved in methanol	173.44 ± 0.49
16. Standardized extract dissolved in ethanol	231.15 ± 0.17

The results for flavonoid content in the *Ginkgo biloba* extracts and infusions (1 – 16) ranged from 14.46 ± 0.23 mg of Ru/g to 231.15 ± 0.17 mg of Ru/g of extract and plant material. Flavonoid contents in the dry extracts obtained by different solvents and evaporation (1 – 7) ranged from 15.94 ± 0.25 mg of Ru/g for petroleum ether extract to 67.65 ± 0.29 mg of Ru/g for ethanolic extract. The results indicate that the amount of flavonoids varied slightly between methanolic (47.70 ± 0.18 mg of Ru/g), buthanolic (47.16 ± 0.44 mg of Ru/g), ethyl acetate (46.34 ± 0.19 mg of Ru/g) and acetone (44.14 ± 0.15 mg of Ru/g) extract. As in the case of the quantity of total phenolics content, type of solvent also influences the amount of flavonoids in the obtained extracts. The obtained results for the flavonoid content in the organic solvent infusions (8 and 9) are 26.66 ± 0.31 mg of Ru/g for methanolic and 33.67 ± 0.28 mg of Ru/g for ethanolic infusion.

Flavonoid contents in the water infusions (10 – 14) obtained using different methods for preparation ranged from 14.46 ± 0.23 mg of Ru/g to 17.92 ± 0.30 mg of Ru/g. On the basis of these results, it can be concluded that the amount of flavonoids does not vary significantly between studied water infusions. As opposed to the quantity of total phenolics content, content of flavonoids depends very little on the methodology of preparation of water infusions. The results for the flavonoid content in the standardized extract of *Ginkgo biloba* leaves (15 – 16) are 173.44 ± 0.49 mg of Ru/g for methanolic and 231.15 ± 0.17 mg of Ru/g for ethanolic solution. The differences in amounts of flavonoids between these extracts point to the importance of the type of the selected solvent. In this case ethanol contributes to better dissolution of flavonoids and their larger amount in the final sample.

After quantification of total phenolics as well as flavonoids separately, as the most important group of phenolic compounds, antioxidant activity of all samples of plant extracts and infusions from *Ginkgo biloba* leaves was determined using free radical (DPPH) assay. The applied method is based on measuring the intensity of reduction of DPPH stable free radicals by active substances in the sample as a hydrogen atom donors. In this process, the purple color of the reaction mixture changes to yellow, where the intensity of the color change, as indicator of antioxidant activity, was quantified spectrophotometrically [58]. The results for the antioxidant activity of analyzed *Ginkgo biloba* plant extracts and infusions expressed in terms of IC_{50} (μg/ml) values are presented in Table 3. In the presentation of results as IC_{50} values, the intensity of antioxidant activity is inversely proportional to the numerical IC_{50} (a lower numerical value indicates better efficiency of plant extracts) values.

The obtained results for the antioxidant activity of *Ginkgo biloba* extract and infusions ranged from 1408. 96 ± 2.01 μg/ml to 49.75 ± 0.90 μg/ml. The antioxidant activity of the extracts obtained by different solvents and evaporation (1 – 7) ranged from 1408. 96 ± 2.01 μg/ml for buthanolic extract to 280.76 ± 1.56 mg of μg/ml for ethanolic extract. The obtained results for the antioxidant activity of the organic solvent infusions (8 and 9) are 322.72 ± 1.04 mg μg/ml for methanolic and 164.31 ± 1.32 μg/ml for ethanolic infusion. The results for the antioxidant activity of the water infusions (10 – 14) obtained using different methods for preparation ranged from 361.15 ± 0.88 μg/ml to 170.12 ± 1.72 μg/ml. The results for the antioxidant activity of the standardized extract of *Ginkgo biloba* leaves (15 – 16) are 62.65 ± 0.84 μg/ml for methanolic and 49.75 ± 0.90 μg/ml for ethanolic solution.

Table 3. Antioxidant activity of the Ginkgo biloba leaves extracts and infusions expressed in terms of IC50 values (μg/ml)

Type of extract	IC_{50} (μg/ml)
1. Water extract	338.72 ± 1.12
2. Methanolic extract	387.13 ± 0.95
3. Ethanolic extract	280.76 ± 1.56
4. Acetone extract	748.22 ± 1.22
5. Ethyl acetate	1212.15 ± 1.36
6. Buthanolic extract	1408. 96 ± 2.01
7. Petroleum ether	1025.70 ± 2.24
8. Methanolic infusion	322.72 ± 1.04
9. Ethanolic infusion	164.31 ± 1.32
10. Water infusion (10°C, 24 h)	361.15 ± 0.88
11. Water infusion (50°C, 24 h)	272.90 ± 1.61
12. Water infusion (100°C, 24 h)	222.97 ± 1.44
13. Water infusion (50°C, 1 h)	177.38 ± 1.80
14. Water infusion (100°C, 1 h)	170.12 ± 1.72
15. Standardized extract dissolved in methanol	62.65 ± 0.84
16. Standardized extract dissolved in ethanol	49.75 ± 0.90

The results for the antioxidant activity and the amount of total phenolics content and flavonoids varied according to the type of solvent used, as well as the method of preparation of the extract and infusion. In addition, it can be concluded that there is a relation between the quantity of phenolics, flavonoids and antioxidant activity.

The variability in the total phenolics content in extracts and infusions prepared using the different solvents could be the result of the varying solubility of the phenolic compounds; this variation in solubility may be driven by the solvent polarity. Some studies showed that ethanol and methanol were better extraction solvents for phenolic compounds from plant materials than less polar solvents such as acetone, petroleum ether etc. According to another study, a less polar solvent such as acetone could extract more phenolic compounds from the flowers than more polar solvents, including methanol and water. These differences may be due to the types of phenolic compounds in plant materials. In general, a good balance in polarity is needed in extracting phenolics from plant sources. During the study of quantitative and qualitative characteristic of flavonoids from plant material, it was found that ethanol and solvents with similar polarity are very effective in the extraction process as well as extracts obtained using this solvent containing a large amount of flavonoids [24, 26, 59].

The investigated extracts and infusions of *Ginkgo biloba* leaves demonstrated very different radical-scavenging activities and IC_{50} values varied in a wide range. The ethanolic extract and infusions showed the greatest effect among the others, and exhibited the greatest radical-scavenging activity. In addition, the phenolic content, as well as flavonoids of extracts and infusions depend on the solvent used in the experiment, and not only the concentration of phenolics but also the properties of these compounds contribute to the activites of different extracts. In numerous studies of biologial activity, for components present in plant material of *Ginkgo biloba* such as lutein [60], luteolin [61], α-carotene [62], cymene [63], thymol [64], zeaxanthin [65], kaempferol [66], quercetin [67], myricetin [68], apigenin [69], as well as catechin and tannins [70] antioxidant activity was determined using different methodological approaches.

Comparing the concentration of phenolic compounds and values for antioxidant activity we found that extracts and infusions with the highest concentrations of phenolic compounds and flavonoids also have strong scavenging effect. Based on these results, extracts and infusions of *Ginkgo biloba* exhibited phenolic concentration-dependent scavenging effects. Numerous investigations of the antioxidant activity of plant extracts have confirmed a high linear correlation between the values of phenolic content and antioxidant activity [71].

CONCLUSION

In this study, the basic profile of total phenolic compounds, total flavonoids and biological activity by measuring of antioxidant capacity of *Ginkgo biloba* leaves extracts and infusions was determined. All parameters were determined for plant extracts obtained using solvents with different polarity (water, methanol, ethanol, acetone, ethyl acetate, buthanol and petroleum ether), infusions obtained using methanol and ethanol, water infusions obtained by different methods for preparations, as well as for standardized *Ginkgo biloba* extract dissolved in methanol and ethanol. Total phenolics content and flavonoids, as well as antioxidant activity varied according to both the type of solvent used and the method of preparation of the extract and infusion. The comparison of the effectiveness of various solvents and methods for infusion preparation showed large variability. Polar solvents such as ethanol and boiled water have the highest extraction efficiency. The ethanolic crude extract, ethanolic infusion, water infusion obtained using boiled water from powdered *Ginkgo biloba* leaves and ethanolic solution of standardized extract contain the greatest concentrations of phenolic compounds, especially flavonoids and showed high antioxidant activity. The comparative analysis indicates that the amount of phenolic compounds and their activity depends on the solvent and method (water temperature) used for extraction and infusion preparation. The results also suggest that there is a relation between the quantity of phenolics, flavonoids and antioxidant activity. A high value of antioxidant activity of *Ginkgo biloba* crude leaves extracts and infusions was proven by comparing with the obtained results for *Ginkgo biloba* standardized extract. The results of this study suggest that *Ginkgo biloba* leaves have high concentrations of phenolic compounds which have quite noticeable effects on the scavenging of free radicals. The extracts of these plant species can be regarded as promising candidates for a natural source of biologically active substances. Comparative analysis of different extracts and infusions obtained using different methods as well as the selection of effective solvent and method for extract and infusion preparation can be helpful when estimating the beneficial properties of *Ginkgo biloba* leaves as valuable medicinal raw plant materials to be used as natural antioxidants in phytopharmacy.

REFERENCES

[1] Pichersky, E; Gang, DR. (2000). Genetics and biochemistry of sccondary metabolites in plants: an evolutionary perspective. *Trends in Plant Science Perspectives*, 5 (10), 439-445.

[2] Edreva, A; Velikova, V; Tonsev, T; Dagnon, S; Gurel, A; Aktas, L; Gesheva, E. (2008). Metabolites: diversity of functions and mechanisms. *General and Applied Plant Physiology*, 34 (1-2), 67-78.

[3] Wink, M. (2004). Phytochemical diversity of secondary metabolites. *Encyclopedia of Plant and Crop Science*, 915-919.

[4] Mazid, M; Khan, TA; Mohammad, F. (2011). Role of secondary metabolites in defense mechanism of plants. *Biology and Medicine*, 3 (2), 232-249.

[5] Shetty, K; Wahlqvist, M. (2004). A model for the role of the proline-linked pentose-phosphate pathway in phenolic phytochemical bio-

synthesis and mechanism of action for human health and environmental applications. *Asia Pacific Journal of Clinical Nutrition*, 13 (1), 1-24.

[6] Quideau, S; Deffieux, D; Douat-Casassus, C; Pouysegu, L. (2011). Plant polyphenols: chemical properties, biological activities, and synthesis. *Natural Products*, 50, 586-621.

[7] Korkina, LG. (2007). Phenylpropanoids as naturally occurring antioxidants: from plant defence to human health. *Cellular and Molecular Biology*, 53 (1), 15-25.

[8] Savithramma, N; Rao, ML; Suhrulatha, D. (2011). Screening of medicinal plants for secondary metabolites. *Middle-East Journal of Scientific Research*, 8 (3), 579-584.

[9] Briskin, DP. (2000). Medicinal plants and phytomedicines, linking plant biochemistry and physiology to human health. *Plant Physiology*, 124, 507-514.

[10] Katalinic, V; Milos, M; Kulisic, T; Jukic, M. (2005). Screening of 70 medicinal plant extracts for antioxidant capacity and total phenols. *Food Chemistry,* 94, 550-557.

[11] Gordaliza, M. (2007). Natural products as leads to anticancer drugs. *Clinical and Translational Oncology*, 9 (12), 767-776.

[12] Hasani-Ranjbar, S; Nayebi, N; Larijani, B; Abdollahi, M. (2010). A systematic review of the efficacy and safety of *Teucrium* species; from anti-oxidant to anti-diabetic effects. *International Journal of Pharmacology,* 6 (4), 315-325.

[13] Karsha, PV; Lakshmi, BO. (2010). Antibacterial activity of black pepper (*Piper nigrum* L.) with special reference to its mode of action on bacteria. *Indian Journal of Natural Products and Resources*, 1 (2), 213-215.

[14] Sen, S; Chakraborty, R; Sridhar, C; Reddy, YSR; De, B. (2010). Free radicals, antioxidants, diseases and phytomedicines: current status and future prospect. *Internacional Journal od Pharmaceutical sciences Review and Research*, 3 (1), 91-100.

[15] Halliwell, B; Whiteman, M. (2004). Measuring reactive species and oxidative damage *in vivo* and in cell culture: how should you do it and what do the results mean? *British Journal of Pharmacology*, 142 (2), 231-255.

[16] Anderson, D. (1996). Antioxidant defences against reactive oxygen species causing genetic and other damage. *Mutation Research*, 350, 103-108.

[17] Valko, M; Rhodes, CJ; Moncol, J; Izakovic, M; Mazur, M. (2006). Free radicals, metals and antioxidants in oxidative stress-induced cancer. *Chemico-Biological Interactions*,160, 1-40.

[18] Devasagayam, TPA; Boloor, KK; Sane, KS; Ghaskadbi, SS; Lele, RD. (2004). Free radicals and antioxidants in human health: current status and future prospect. *Journal of the Association of Physicians of India,* 52, 794-804.

[19] Gupta, VK; Sharma, SK. (2006). Plants as natural antioxidants. *Natural Product Radiance,* 5 (4), 326-334.

[20] Ito, N; Hirose, M; Fukushima, H; Tsuda, T; Shirai, T; Tatenatsu, M. (1986). Studies on antioxidants: Their carcinogenic and modifying effects on chemical carcinogens. *Food and Chemical Toxycology*, 24, 1071 1092.

[21] Chen, C; Pearson, MA; Gray, IJ. (1992). Effects of synthetic antioxidants (BHA, BHT and PG) on the mutagenicity of IQ-like compounds. *Food Chemistry*, 43, 177-183.

[22] Kahl, R; Kappus, H. (1993). Toxicology of the synthetic antioxidants BHA and BHT in comparison with the natural antioxidant vitamin E. *Zeitschrift für Lebensmittel-Untersuchung und-Forschung A,* 196 (4), 329-338.

[23] Beer, D; Joubert, E; Gelderblom, WCA; Manley, M. (2002). Phenolic compounds: a review of their possible role as *in vivo* antioxidants. *South African Journal of Enology and Viticulture*, 23 (2), 48-61.

[24] Dai, J; Mumper, RJ. (2010). Plant phenolics: extraction, analysis and their antioxidant and anticancer properties. *Molecules,* 15, 7313-7352.

[25] Stanković, SM; Topuzović, M; Solujić, S; Mihailović, V. (2010). Antioxidant activity and concentration of phenols and flavonoids in the whole plant and plant parts of *Teucrium chamaedrys* L. var. *glanduliferum* Haussk. *Journal of Medicinal Plants Research*, 4 (20), 2092-2098.

[26] Stanković, SM; Nićiforović, N; Mihailović, V; Topuzović, M; Solujić, S. (2012). Antioxidant activity, total phenolic content and flavonoid concentrations of different plant parts of *Teucrium polium* L. subsp. *polium*. *Acta Societatis Botanicorum Poloniae*, 81 (2), 117-122.

[27] Del Tredici, P. (2000). The evolution, ecology, and cultivation of *Ginkgo biloba*. In: Van Beek ART. *Ginkgo biloba*. Amsterdam, The Netherlands, Harwood Academic Publishers, 7-23.

[28] Singh, B; Kaur, P; Gopichand, Singh, RD; Ahuja, PS. (2008). Biology and chemistry of *Ginkgo biloba*. *Fitoterapia,* 79 (6), 401-418.

[29] Van Beek, TA. (2002). Chemical analysis of *Ginkgo biloba* leaves and extracts. *Journal of Chromatography A*, 967 (1), 21-55.

[30] Ude, C; Schubert-Zsilavecz, M; Wurglics, M. (2013). *Ginkgo biloba* extracts: A review of the pharmacokinetics of the active ingredients. *Clinical Pharmacokinetics*, 52 (9), 727-749.

[31] Kleijnen, J; Knipschild, P. (1992). *Ginkgo biloba* for cerebral insufficiency. *British Journal of Clinical Pharmacology*, 34, 352-358.

[32] Pietri, S; Maurelli, E; Drieu, K; Culcasi, M. (1997). Cardioprotective and anti-oxidant effects of the terpenoid constituents of *Ginkgo biloba* extract (EGb 761). *Journal of Molecular and Cellular Cardiology*, 29, 733-742.

[33] Ernst, E; Pittler, MH. (1999). A systematic review of double-blind, placebo-controlled trials *Ginkgo biloba* for dementia. *Clinical Drug Investigation,* 17 (4), 301-308.

[34] Watanabe, CM; Wolffram, S; Ader, P; Rimbach, G; Packer, L; Maguire, JJ; Schultz, GP; Gohil, K. (2001). The *in vivo* neuromodulatory effects of the herbal medicine *Ginkgo biloba*. *Proceedings of the National Academy of Sciences*, 98 (12), 6577-6580.

[35] Zimmermann, M; Colciago, F; Cattabeni, F; Di Luca, M. (2002). *Ginkgo biloba* extract: from molecular mechanisms to the treatment of Alzheimer's disease. *Cellular and Molecular Biology*, 48, 613-623.

[36] Ahlemeyer, B; Krieglstein, J. (2003). Neuroprotective effects of *Ginkgo biloba* extract. *Cellular and Molecular Life Sciences*, 60, 1779-1792.

[37] Van Dongen, M; Van Rossum, E; Kessels, A; Sielhorst, H; Knipschild, P. (2003). *Ginkgo* for elderly people with dementia and age-associated memory impairment: a randomized clinical trial. *Journal of Clinical Epidemiology,* 56, 367-376.

[38] Smith, JV; Luo, Y. (2004). Studies on molecular mechanisms of *Ginkgo biloba* extract. *Applied Microbiology and Biotechnology*, 64, 465-472.

[39] Walesiuk, A; Trofimiuk, E; Braszko, JJ. (2005). *Ginkgo biloba* extract diminishes stress-induced memory deficits in rats. *Pharmacological Reports,* 57, 176-187.

[40] Mahadevan, S; Park, Y. (2008). Multifaceted therapeutic benefits of *Ginkgo biloba* L.: chemistry, efficacy, safety, and uses. *Journal of Food Science,* 73 (1), R14-R19.

[41] Mazzanti, G; Mascellino, MT; Battinelli, L; Coluccia, D; Manganaro, M; Saso, L. (2000). Antimicrobial investigation of semi purified fractions of *Ginkgo biloba* leaves. *Journal of Ethnopharmacology*, 71, 83-88.

[42] Park, SB; Cho, GS. (2011). Antimicrobial activity of extracts and fractions of *Ginkgo biloba* leaves, seed and outer seed coat. *Journal of the Korean Society of Food Science and Nutrition*, 40 (1), 7-13.

[43] Sati, SC; Joshi, S. (2011). Antibacterial activities of *Ginkgo biloba* L. leaf extracts. *The Scientific World Journal*, 11, 2237-2242.

[44] Tao, R; Wang, CZ; Ye, JZ; Zhou, H; Chen, H; Zhang, Y. (2014). Antibacterial/antifungal activity and synergistic interactions between C_{70}-C_{120} polyprenol homologs from *Ginkgo biloba* L. leaves and the corresponding synthetic derivatives. *European Food Research and Technology,* 239 (4), 587-594.

[45] Bajpai KV; Sharma, A; Kwang-Hyun, B. (2015). Antibacterial mode of action of *Ginkgo biloba* leaf essential oil: Effect on morphology and membrane permeability. *Bangladesh Journal of Pharmacology*, 10 (2), 337-350.

[46] Naik, SR; Pilgaonkar, VW; Panda, VS. (2006). Evaluation of antioxidant activity of *Ginkgo biloba* phytosomes in rat brain. *Phytotherapy Research,* 20 (11), 1013-1016.

[47] Naik, SR; Panda, VS. (2007). Antioxidant and hepatoprotective effects of *Ginkgo biloba* phytosomes in carbon tetrachloride-induced liver injury in rodents. *Liver International*, 27, 393-399.

[48] Kaur, A; Nain, P; Nain, J. (2012). *In vitro* antioxidant and antimicrobial activity of *Ginkgo biloba* bark extract. *International Journal of Natural Products Science*, 1, 168.

[49] Krizková, L; Chovanová, Z; Duracková, Z; Krajcovic, J. (2008). Antimutagenic *in vitro* activity of plant polyphenols: Pycnogenol and *Ginkgo biloba* extract (EGb 761). *Phytotherapy Research*, 22 (3), 384-388.

[50] Vilar, BJ; Leite, RK; Chen Chen, L. (2009). Antimutagenicity protection of *Ginkgo biloba* extract (Egb 761) against mitomycin C and cyclophosphamide in mouse bone marrow. *Genetics and Molecular Research,* 8 (1), 328-333.

[51] Feng, X; Zhang, L; Zhu, H. (2009). Comparative anticancer and antioxidant activities of different ingredients of *Ginkgo biloba* extract (EGb 761). *Planta Medica,* 75 (8), 792-796.

[52] Jiang, W; Qiu, W; Wang, Y; Cong, Q; Edwards, D; Ye, B; Xu, C. (2011). *Ginkgo* may prevent genetic-associated ovarian cancer risk: multiple biomarkers and anticancer pathways induced by ginkgolide B in BRCA1-mutant ovarian epithelial cells. *European Journal of Cancer Prevention*, 20 (6), 508-517.

[53] Zhou, ZY; Tang, SQ; Zhou, YM; Luo, HS; Liu, X. (2010). Antioxidant and hepatoprotective effects of extract of *Ginkgo biloba* in rats of non-alcoholic steatohepatitis. *Saudi Medical Journal*, 31 (10), 1114-1118.

[54] Singleton, VL; Orthofer, R; Lamuela, RRM. (1999). Analysis of total phenols and other oxidation substrates and antioxidants by means of Folin-Ciocalteu reagent. *Methods in Enzymology*, 299, 152-178.

[55] Quettier, DC; Gressier, B; Vasseur, J; Dine, T; Brunet, C; Luyckx, MC; Cayin, JC; Bailleul, F; Trotin, F. (2000). Phenolic compounds and antioxidant activities of buckwheat (*F. esculentum* Moench) hulls and flour. *Journal of Ethnopharmacology*, 72 (1-2), 35-42.

[56] Tekao, T; Watanabe, N; Yagi, I; Sakata, K. (1994). A simple screening method for antioxidant and isolation of several antioxidants produced by marine bacteria from fish and shellfish. *Bioscience, Biotechnology and Biochemistry,* 58, 1780-1783.

[57] Kumarasamy, Y; Byres, M; Cox, PJ; Jasapars, M; Nahar, L; Sarker, SD. (2007). Screening seeds of some Scottish plants for free-radical scavenging activity. *Phytotherapy Research*, 21, 615-621.

[58] Xie, J; Schaich, KM. (2014). Re-evaluation of the 2,2-Diphenyl-1-picrylhydrazyl free radical (DPPH) assay for antioxidant activity. *Journal of Agricultural and Food Chemistry*, 62, 4251-4260.

[59] Mandana, B; Russly, AR; Farah, ST; Ali, G; Liza, MS; Jinap, S; Azizah, H. (2011). Comparison of different extraction methods for the extraction of major bioactive flavonoid compounds from spearmint (*Mentha spicata* L.) leaves. *Food and Bioproducts Processing*, 89, 67-72.

[60] Sindhu, ER; Preethi, KC; Kuttan, R. (2010). Antioxidant activity of carotenoid lutein *in vitro* and *in vivo*. *Indian Journal of Experimental Biology,* 48 (8), 843-848.

[61] Xu, K; Liu, B; Ma, Y; Du, J; Li, G; Gao, H; Zhang, Y; Ning, Z. (2009). Physicochemical properties and antioxidant activities of luteolin-phospholipid complex. *Molecules,* 14, 3486-3493.

[62] Farombi, EO; Britton, G. (1999). Antioxidant activity of palm oil carotenes in peroxyl radical-mediated peroxidation of phosphatidyl choline liposomes. *Redox Report,* 4 (1-2), 61-68.

[63] De Oliveira, TM; De Carvalho, RB; De Costa, IH; De Oliveira, GA; De Souza, AA; De Lima, SG; De Freitas, RM. (2015). Evaluation of p-cymene, a natural antioxidant. *Pharmaceutical Biology*, 53 (3), 423-428.

[64] Kumar, BD; Rawat, SD. (2013). Synthesis and antioxidant activity of thymol and carvacrol based Schiff bases. Bioorganic and Medicinal *Chemistry Letters*, 23 (3), 641-645.

[65] Santocono, M; Zurria, M; Paladino, G. (2003). Antioxidant activity of the xanthophylles astaxanthin, lutein and zeaxanthin: *In vitro* assays. *Investigative Ophthalmology and Visual Science*, 44, 1699.

[66] Dar, AR; Brahmana, KP; Khurana, N; Wagaye, AJ; Lone, AZ; Ganaie, AM; Pitre, SK. (2013). Evaluation of antioxidant activity of crocin, podophyllotoxin and kaempferol by chemical, biochemical and electrochemical assays. *Arabian Journal of Chemistry,* doi:10.1016/j.arabjc.2013.02.004.

[67] Dueñas, M; Surco-Laos, F; González-Manzano, S; González-Paramás, AM; Santos-Buelga, C. (2011). Antioxidant properties of major metabolites of quercetin. *European Food Research and Technology*, 232 (1), 103-111.

[68] Bryan, RM. Efficacy of myricetin as an antioxidant in methyl esters of soybean oil. *European Journal of Lipid Science and Technology*, 110 (12), 1167-1174.

[69] Li, J. (2014). Study and characterization of the antioxidant activity of the inclusion complex of apigenin with *β*-cyclodextrin and HP-*β*-cyclodextrin in solution. *Journal of Investigational Biochemistry*, 3 (3), 107-111.

[70] Dangles, O; Fargeixa, G; Dufourb, C. (2000). Antioxidant properties of anthocyanins and tannins: a mechanistic investigation with catechin and the 3′,4′,7-trihydroxyflavylium ion. *Journal of the Chemical Society,* Perkin Transactions, 2, 1653-1663.

[71] Leccese, A; Viti, R; Bartolini, S. (2011). The effect of solvent extraction on antioxidant properties of apricot fruit. *Central European Journal of Biology,* 6, 199-204.

BIBLIOGRAPHY

Complementary therapies for physical therapy: a clinical decision-making approach LCCN: 2010455083 Personal name: Deutsch, Judith E. (Judith Erica), 1959- Main title: Complementary therapies for physical therapy: a clinical decision-making approach / Judith E. Deutsch, Ellen Zambo Andcrson. Published/Created: St. Louis, Mo.: Saunders/Elsevier, c2008. Description: xxi, 327 p.: ill.; 28 cm. ISBN: 0721601111 9780721601113 LC classification: RM701 .D48 2008 Related names: Anderson, Ellen Zambo. Contents: ch. 1 CAM use in illness and wellness / Judith E. Deutsch -- ch. 2 Conceptual framework for clinical decision making in complementary and alternative medicine / Judith E. Deutsch, Ellen Zambo Anderson -- ch. 3 Modifiers of complementary therapy: legal, ethical, and cultural issues / Ellen Zambo Anderson -- ch. 4 Whole medical systems / Judith E. Deutsch, Suzanne McDonough -- ch. 5 Acupuncture / Suzanne McDonough, Sheelagh McNeill -- ch. 6 Arnica / Lori Zucker -- ch. 7 Overview of mind-body therapies / Susan Gould Fogerite, Gary L. Goldberg -- ch. 8 Yoga / Mary Lou Galantino, John Musser -- ch. 9 Tai chi / Patricia Quinn McGinnis -- ch. 10 Overview of biologically based therapies in rehabilitation / Susan Gerik, John Maypole -- ch. 11 Ginkgo biloba / Ellen D. Mandel -- ch. 12 Glucosamine chondroitin / Diane Rigassio Radler -- ch. 13 Energy therapy / Ellen Zambo Anderson -- ch. 14 Therapeutic touch / Ellen Zambo Anderson -- ch. 15 Qigong / Bill Gallagher,

Richard Lund -- ch. 16 Magnets / Ellen Zambo Anderson, Cathy Caro-Scarpito -- ch. 17 Reiki / Ellen Zambo Anderson, Cindy Wolk-Weiss -- ch. 18 Manipulative and body-based therapies / Judith E. Deutsch -- ch. 19 The Ida Rolf method of structural integration / Judith E. Deutsch -- ch. 20 Feldenkrasis / James Stephens -- ch. 21 The Alexander technique / Glenna Batson -- ch. 22 Craniosacral therapy / Ellen Zambo Anderson, Perry Wolk-Weiss -- ch. 23 Pilates / Ellen Zambo Anderson, Chantel Dickinson. Subjects: Physical therapy. Alternative medicine. Medical rehabilitation. Complementary Therapies. Physical Therapy Modalities. Notes: Includes bibliographical references and index.

Evidence and rational based research on Chinese drugs LCCN: 2012955141 Main title: Evidence and rational based research on Chinese drugs / Hildebert Wagner, Gudrun Ulrich-Merzenich, editors. Published/Produced: Wien; New York: Springer, [2013] ©2013 Description: xii, 525 pages: illustrations; 24 cm ISBN: 9783709104415 (alk. paper) 3709104416 (alk. paper) LC classification: RM301.25 .E94 2013 Related names: Wagner, Hildebert, 1929- editor. Ulrich-Merzenich, Gudrun, editor. Summary: After the successful introduction of acupuncture to the West, recent advances in analytic methods in chemistry, molecular biology and system biology - especially the development of the "omic"-technologies -- have again brought Chinese drugs into the focus of research in traditional Chinese medicine (TCM). With more than 1000 publications on the chemistry, molecular biology and pharmacology of TCM drugs in international journals over the last 10 years, Chinese drugs are gaining reputation and impact. These data offer great opportunities for the development of new pharmaceuticals for various clinical applications. Scientist from Europe, USA and China are presently compiling the relevant and trend setting research results in a book. Topics range from the latest methods of quality and safety proof by chemical and genetic fingerprints to the development of new pharmaceuticals for a future evidence-based therapy e.g., for cancer, cardiovascular, inflammatory or infectious diseases. Recent experimental results on multitarget and synergy research for the preparation of multi-extract-pharmaceuticals from TCM are equally covered.-- Source other than Library of Congress. Contents: Development of New Analytical Monographs of Herbal Drugs from TCM for Quality Proof and Development of New Phytopharmaceuticals / H. Wagner -- DNA-Based Authentication of TCM-Plants: Current Progress and Future Perspectives / G. Heubl -- Newest Results on the Chemistry and Pharmacology of TCM Drugs Containing Triterpene and Steroid Saponins / Marie-Aleth Lacaille-Dubois -- Efficacy of Andrographis paniculata in Upper Respiratory Tract Infectious Diseases and the Mechanism of Action / Alexander Panossian, Georg Wikman -- From Traditional to Evidence-Based Use of Hippophae rhamnoides L.: Chemical Composition, Experimental, and Clinical Pharmacology of Sea Buckthorn Berries and Leaves Extracts / Alexander Panossian, Hildebert Wagner -- New Results on the Pharmacology and Clinical Use of the TCM-Drug Salvia miltiorrhiza / John H. K. Yeung -- Inhibition of ATP-Binding Cassette Transporters by Chinese Herbs and Phytochemicals / Thomas Efferth -- Activity of Artemisinin-Type Compounds Against Cancer Cells / Serkan Sertel, Peter K. Plinkert -- Chinese Herbal Medicines for Neuroprotection in Ischemic Stroke: Promise and Reality / Nikolaus J. Sucher -- Complementary and Traditional Chinese Medicine Methods in the Treatment of Gynecological Diseases / Wolfgang Wuttke, Dana Seidlova-Wuttke -- Ginkgo biloba Extract EGb 761®: From an Ancient Asian Plant to a Modern European Herbal Medicinal Product / Friedrich Lang, Robert Hoerr -- Ginkgolides and Their Derivatives: Synthetic and Bioorganic Studies / Sergei V. Dzyuba, Laramie P. Jameson -- Towards a Contemporary and Evidence-Based Development of TCM / Hildebert Wagner, Gudrun Ulrich-Merzenich. Subjects: Drugs--Research. Drugs--Testing. Medicine, Chinese--Formulae, receipts, prescriptions. Drugs, Chinese Herbal--pharmacology. Drug Discovery. Drugs, Chinese Herbal--analysis. Medicine, Chinese Traditional. Médecine traditionnelle chinoise. Drugs--Research. Drugs--Testing. Medicine, Chinese. Pharmacologie--Chine.

Médecine chinoise. Form/Genre: Prescriptions, formulae, receipts, etc. Notes: Includes bibliographical references and index.

Handbook on flavonoids: dietary sources, properties, and health benefits LCCN: 2011041379 Main title: Handbook on flavonoids: dietary sources, properties, and health benefits / editors, Kazuya Yamane and Yuudai Kato. Published/Created: Hauppauge, N.Y.: Nova Science Publishers, c2012. Description: xv, 557 p.: ill.; 27 cm. ISBN: 9781619420496 (hardcover) LC classification: QP671.F52 H36 2012 Related names: Yamane, Kazuya. Kato, Yuudai. Contents: Flavonoids: recent insights on their biological action / Salvatore Chirumbolo -- Pharmacokinetic variability of dietary phenolic acids and flavonoids in relation to chemical and biological factors / Nabil Semmar, Asma Hammami-Semmar -- Modification of flavonoid structure by oxovanadium (IV) complexation: biological effects / Evelina G. Ferrer, Patricia A.M. Williams -- Flavonoids and its contribution to a healthier life / Maria do Rosário Bronze, Maria Eduardo Figueira, Elsa Mecha -- Effects of some domestic cooking methods on antioxidant activity, flavonoids, and other phytochemicals content / Irene Dini -- Health effects on flavonoids and their relationship in mushrooms / Noboru Motohashi -- Dietary flavonoids modulate the oxidative DNA damage induced by N-nitrosamines, heterocyclic amines, and benzo(a)pyrene / Paloma Morales, Ana I. Haza -- Impact of conventional and non-conventional technologies applied to obtain fruit products in the flavonoid content and antioxidant capacity of grapefruit / M. Igual ... [et al.] -- UV-B radiation: a powerful tool to modulate flavonoid metabolism in tomato fruits / Annamaria Ranieri -- Anti-inflammatory properties of dietary flavonoids / A.García-Lafuente, E. Guillamón -- Flavonoids: from food and its implication in human health / Montse Rabassa ... [et al.] -- Processing of citrus peel for the extraction of flavonoids for biotechnological applications / Munish Puri, Madan Lal Verma, Kiran Mahale -- Regulation of intestinal barrier function by dietary flavonoids / Takuya Suzuki -- Anti-cancer mechanisms of flavonoids in malignant neuroblastoma / Mrinmay Chakrabarti, Swapan

K. Ray -- Flavonoid distribution in neglected citrus species grown in the Mediterranean basin / Davide Barreca ... [et al.] -- Flavonoids in mushrooms: occurrence, properties, and role of their antioxidant activity / A. Villares -- Ginkgo biloba leaves extract (EGb 761) and its specific acylated flavonol constituents increase dopamine and acetylcholine levels in the rat medial prefrontal cortex: possible implications for cognitive enhancing properties of the ginkgo extract / Takashi Yoshitake ... [et al.] -- Dietary sources of isoflavones and the methodology used for the analysis / Savithiry S. Natarajan, Devanand L. Luthria . Subjects: Flavonoids. Notes: Includes bibliographical references and index.

Herbal radiomodulators: applications in medicine, homeland defence and space LCCN: 2008005764 Main title: Herbal radiomodulators: applications in medicine, homeland defence and space / editor, Rajesh Arora. Published/Created: Wallingford, UK; Cambridge, MA: CABI, c2008. Description: xvii, 332 p.: ill.; 26 cm. Links: Table of contents only http://www.loc.gov/catdir/toc/ecip0810/2008005764.html ISBN: 1845933958 (hbk.: alk. paper) 9781845933951 (hbk.: alk. paper) LC classification: RM849 .H47 2008 Related names: Arora, Rajesh. Contents: Radiomodulatory compounds of herbal origin for new frontiers in medicine, homeland security, management of radiological incidents, and space applications / Rajesh Arora ... [et al.] -- Indian medicinal herbs and ayurvedic formulations as potential radioprotectors / D.K. Maurya and T.P.A. Devasagayam -- Irradiation, radioprotection, and Nigella sativa / M. Cemek ... [et al.] -- Modulation of radiation-induced damage by Serbian natural plant products: implications for radioprotection / Gordana Joksić, Andreja Leskovac, and Sandra Petrović -- Phytoceuticals for radioprotection with special reference to Egyptian flora / N.M. Abdel-Hamid -- Melatonin mitigates the damaging effects of ionizing radiation / Russel J. Reiter ... [et al.] -- Radioprotective effect of citrus and hawthorn extracts against genotoxicty induced by gamma irradiation / Seyed Jalal Hosseinimehr -- The healing potential of indigenous essential oils from New Zealand in the prevention and management of radiation-induced mucositis / W. Maddocks-Jennings -- Piper betel leaves: a potential gold mine of radioprotective and photoprotective compounds / Debashish Banerjee and Subrata Chattopadhyay -- Dietary antioxidants and phytochemicals in radioprotection and therapy / Carmia Borek -- Effects of berry fruits on neurocognitive deficits produced by exposure to space radiation / B.M. Rabin, James Joseph, and Barbara Shukitt-Hale -- Radioprotection by the soy isoflavone genistein / Michael R. Landauer -- Propolis and related flavonoids as radioprotective agents / Nada Oršolić ... [et al.] -- Radioprotective effects of Ginkgo biloba via its antioxidant action / Göksel Şener, Abdullah Sakarcan, and Berrak Ç. Yeğen -- Novel strategies for protecting mitochondria (the cellular powerhouse) against low-LET radiation: a review / Damodar Gupta ... [et al.] -- Andrographis paniculata: an emerging radioprotective agent for membrane proteins / Rakshamani Tripathi and Jayashree P. Kamat -- Mitigation of deleterious effects of ionizing radiation by phytoceuticals: mechanistic studies with Centella asiatica / C.K.K. Nair and Jisha Joy -- The radiosensitizing effects of L-canavanine / David R. Worthen and Peter A. Crooks -- Withaferin A: a phytosteroid of promise for tumour sensitization in cancer therapy / P. Uma Devi -- The radiosensitizer hypericin as adjuvant therapy in the treatment of central nervous system tumours / Toba Niazi and William T. Couldwell -- Radiosensitizing activity of the Indian medicinal plant Tinospora cordifolia Miers ex Hook F and Thoms in tumour-bearing mice / Ganesh Chandra Jagetia -- Do antioxidants reduce the efficacy of radiotherapy? / Ralph Moss. Subjects: Radiation-protective agents. Herbs--Therapeutic use. Radiotherapy. Radiation-Protective Agents--pharmacology. Radiation-Protective Agents--therapeutic use. Phytotherapy. Plant Preparations--therapeutic use. Plants, Medicinal. Radiation Injuries--drug therapy. Notes: Includes bibliographical references and index.

Herbal supplements: efficacy, toxicity, interactions with western drugs, and effects on clinical laboratory tests LCCN: 2010019504 Main title: Herbal supplements: efficacy, toxicity, interactions with western drugs, and effects on clinical laboratory tests / edited by Amitava Dasgupta, Catherine A. Hammett-Stabler. Published/Created: Hoboken, N.J.: John Wiley and Sons, c2011. Description: xiv, 470 p.: ill.; 24 cm. ISBN: 9780470433508 (cloth) LC classification: RA1250 .H47 2011 Related names: Dasgupta, Amitava, 1958- Hammett-Stabler, Catherine A., 1952- Contents: An introduction to complementary and alternative medicine / Catherine A. Hammett-Stabler -- Relatively safe herbal remedies / Angela M. Ferguson and Uttam Garg -- Risk of toxicity associated with unregulated herbal products / Steven W. Cotten -- Herbal medicines with immunomodulatory effects / Jeffrey K. Actor -- Kelp and thyroid function / Bruce Rosenzweig -- Herbal remedies and the patient with chronic kidney disease / Mariana S. Markell -- Abnormal liver function tests due to hepatotoxic herbs / Amitava Dasgupta and Catherine A. Hammett-Stabler -- Homeopathic medicine: principle, efficacy and toxicity / Amitava Dasgupta -- Indian ayurvedic medicines: an introduction / Amitava Dasgupta -- Tradition and perspectives of Greco-Arab and Islamic herbal medicine / Bashar Saad and Omar Said -- Licorice and laboratory tests / Salvador F. Sena -- Drug interactions with St. John's wort / Matthew D. Krasowski and John L. Blau -- Drug-herb interactions in patients with HIV/AIDS / Natella Y. Rakhmanina and John N. van den Anker -- Interactions between fruit juices and drugs / Amitava Dasgupta -- Drug interactions with ginseng and ginkgo biloba / Ashok Tholpady and Semyon A. Risin -- Drug interactions with garlic and ginger supplements / Charbel Abou-Diwan and James Ritchie -- Heavy metal toxicity and herbal remedies / Christine L.H. Snozek and Loralie J. Langman -- Adulteration of herbal remedies with conventional drugs: role of the clinical laboratory / Uttam Garg and Angela M. Ferguson -- Beyond herbals: an introduction to poisonous plants / Catherine A. Hammett-Stabler -- Interferences of herbal remedies with immunoassays for therapeutic drugs: focus on Digoxin / Amitava Dasgupta -- Role of the clinical laboratory in detecting plant poisoning / Ronald W. McLawhon. Subjects: Herbs--Toxicology. Herbs--Therapeutic use. Drug-herb interactions. Plants, Medicinal--chemistry. Plants, Medicinal--toxicity. Dietary Supplements--toxicity. Herb-Drug Interactions. Pathology, Clinical--methods. Phytotherapy--adverse effects. Notes: Includes bibliographical references and index.

Micronutrients and brain health
LCCN: 2009010251 Main title: Micronutrients and brain health / edited by Lester Packer ... [et al.]. Published/Created: Boca Raton: CRC Press, c2010. Description: xxi, 434 p., [4] p. of plates: ill. (some col.); 27 cm. ISBN: 9781420073515 (hbk.) 1420073516 (hbk.) LC classification: QP356.3 .M535 2010 Related names: Packer, Lester. Summary: "Under the direction of leading experts in oxidative stress, this book addresses cutting-edge areas of research regarding micronutrients and the brain. It discusses identification of brain-specific micronutrients that support function and molecular mechanisms underlying neuroprotectant activity. The book covers age-related metabolic pathways, mitochondrial nutrients, and neurodegeneration. Additional chapters cover flavonoids, cell signaling, and neuronal functions, as well as the role of choline, amino acids, metals, and other micronutrients in brain health and function. The text places a particular emphasis on lipoic acid, which is shown to be a therapeutic agent in neuropathologies."--Publisher's description. Contents: Neuroprotection after cardiac arrest by avoiding acute hyperoxia and by antioxidant genomic postconditioning / Gary Fiskum and Robert E. Rosenthal -- The neuroprotective role of micronutrients in Parkinson's disease / Kristen Malkus, Elpida Tsika, and Harry Ischiropoulos -- Phytoestrogens and brain health / Liqin Zhao and Roberta Diaz Brinton -- Food antioxidants and Alzheimer's disease / Emma Ramiro-Puig ... [et al.] -- Micronutrient antioxidants, cognition, and neuropathology: a longitudinal study in the canine model of human aging / Wycliffe O. Opii and Elizabeth Head -- Excitatory amino acids, S-nitrosylation, and protein misfolding in neurodegenerative disease: protection by memantine and nitromemantine at NMDA-gated channels / Tomohiro Nakamura and Stuart A. Lipton -- Cognitive and behavioral consequences of iron deficiency in women of reproductive age /

Laura E. Murray-Kolb -- Micronutrient needs of the developing brain: priorities and assessment / Anita J. Fuglestad, Sara E. Ramel, and Michael K. Georgieff -- Therapeutics of Alzheimer's disease based on metal bioavailability / Su San Mok and Ashley I. Bush -- Lipoic acid as a novel treatment for mild cognitive impairment and early-stage Alzheimer's disease / Annette Maczurek ... [et al.]. Zinc and the cytoskeleton in neuronal signaling / Gerardo G. Mackenzie and Patricia I. Oteiza -- Tocotrienol neuroprotection: the most potent biological function of all natural forms of vitamin E / Chandan K. Sen, Savita Khanna, and Sashwati Roy -- Fruits, nuts, and brain aging: nutritional interventions targeting age-related neuronal and behavioral deficits / James A. Joseph, Barbara Shukitt-Hale, and Lauren M. Willis -- Modulation of multiple pathways involved in the maintenance of neuronal function by fisetin / Pamela Maher -- Dietary flavonoids as neuroprotective agents / Jeremy P. E. Spencer ... [et al.] -- Actions of bioactive phytochemicals in cell function and Alzheimer's disease pathology / Richard E. Hartman -- Does ginkgo biloba extract exert an effect on Alzheimer's disease progression / Yves Christen -- Green tea polyphenols protect neurons against Alzheimer's disease and Parkinson's disease / Baou Zhao -- Transport of flavonoids into the brain / Paul E. Milbury -- Prevention and treatment of neurodegenerative diseases by spice-derived phytochemicals / Bharat B. Aggarwal, Kuzhuvelil B. Harikumar, and Sanjit Dey -- Neurohormetic properties of the phytochemical resveratrol / Andrea Lisa Holme and Shazib Pervaiz -- Sirtuin and resveratrol / Antoni Camins ... [et al.] -- Acetyl-l-carnitine and ferulic acid action in aging and neurodegenerative diseases / Rena A. Sowell, Christopher D. Aluise, and D. Allan Butterfield -- Evidence required for causal inferences about effects of micronutrient deficiencies during development on brain health: DHA, choline, iron, and vitamin D / Joyce C. McCann and Bruce N. Ames -- Omega-3 fatty acids and brain function in older people / Ricardo Uauy and Alan D. Dangour -- Iron and monoamine oxidase in brain function and dysfunction: development of neuroprotective-neurorescue drugs / Orly Weinreb ... [et al.] --

Antioxidative defense of brain microglial cells / Ralf Dringen and Johannes Hirrlinger -- Branched-chain amino acids and brain metabolism / Radovan Murin and Bernd Hamprecht. Subjects: Brain--Metabolism. Trace elements in nutrition. Nutrition. Neuroprotective agents. Oxidative stress. Micronutrients--therapeutic use. Neuroprotective Agents--therapeutic use. Brain--physiology. Neurodegenerative Diseases--prevention and control. Notes: Includes bibliographical references and index. Series: Oxidative stress and disease; 26 Oxidative stress and disease; 26.

Natural medications for psychiatric disorders: considering the alternatives LCCN: 2008019844 Main title: Natural medications for psychiatric disorders: considering the alternatives / editors, David Mischoulon, Jerrold F. Rosenbaum. Edition: 2nd ed. Published/Created: Philadelphia: Lippincott Williams and Wilkins, c2008. Description: xiv, 366 p.: ill.; 24 cm. Links: Publisher description http://www.loc.gov/catdir/enhancements/fy0834/2008019844-d.html Table of contents only http://www.loc.gov/catdir/enhancements/fy0837/2008019844-t.html ISBN: 9780781767620 0781767628 LC classification: RC480.5 .N285 2008 Related names: Mischoulon, David. Rosenbaum, J. F. (Jerrold F.) Contents: Complementary and alternative medicine in society: an introduction / David Mischoulon and Jerrold F. Rosenbaum -- Evaluating the data: limitations of research and quality assurance issues regarding natural remedies / Maurizio Fava and David Mischoulon -- St. John's wort: a critical evaluation of the evidence for antidepressant effects / Andrew A. Nierenberg, Hannah G. Lund, and David Mischoulon -- Omega-3 fatty acids in mood disorders: a review of neurobiological and clinical actions / Andrew L. Stoll -- One-carbon metabolism and the treatment of depression: roles of S-adenosyl-L-methionine and folate / Jonathan E. Alpert, George I. Papakostas, and David Mischoulon -- Dehydroepiandrosterone in the treatment of neuropsychiatric conditions / Owen M. Wolkowitz ... [et al.] -- Inositol in the treatment of psychiatric disorders / Robert H. Belmaker and Joseph Levine -- Herbal remedies for anxiety and insomnia: kava and valerian / David Mischoulon --

Therapeutic potential of melatonin in sleep and circadian disorders / Irina V. Zhdanova and Leah Friedman -- Ginkgo biloba extract in cognitive disorders / Hakima Amri ... [et al.] -- Black cohosh for relief of climacteric symptoms / Daniel S. Fabricant ... [et al.] -- Chaste tree fruit and premenstrual syndrome / Donna E. Webster ... [et al.] -- Natural products with anti-addictive activities / David Yue-Wei Lee -- Complementary and alternative therapy for weight management / Anne E. Becker ... [et al.] -- Acupuncture for the treatment of psychiatric disorders / Albert Yeung ... [et al.] -- Homeopathy and its applications in psychiatry / Iris R. Bell and Pamela A. Pappas -- Polypharmacy, side effect management, and drug-drug interactions with natural psychotropic medications and acupuncture / David Mischoulon and Christina M. Dording. Subjects: Mental illness--Alternative treatment. Naturopathy. Mental Disorders--drug therapy. Complementary Therapies. Psychotropic Drugs--therapeutic use. Notes: Includes bibliographical references and index.

The woman's heart: an owner's guide LCCN: 2007050589 Personal name: Elefteriades, John A. Main title: The woman's heart: an owner's guide / John A. Elefteriades and Teresa Caulin-Glaser. Published/Created: Amherst, N.Y.: Prometheus Books, 2008. Description: 304 p.: ill.; 23 cm. Links: Table of contents only http://www.loc.gov/catdir/toc/ecip087/2007050589.html Contributor biographical information http://www.loc.gov/catdir/enhancements/fy1309/2007050589-b.html Publisher description http://www.loc.gov/catdir/enhancements/fy1309/2007050589-d.html ISBN: 9781591025627 (pbk.) 1591025621 LC classification: RC672 .E44 2008 Related names: Caulin-Glaser, Teresa. Contents: Acknowledgment -- Preface -- Introduction: The "broken heart" syndrome -- ch. 1. The female heart -- 1. Normal structure and function -- 2. Anatomy and physiology of a woman - how the female heart differs -- 3. Changes associated with pregnancy -- ch. 2. Diseases of the heart -- 1. Vessels - coronary artery disease -- 2. Valves - mitral valve disease (including mitral valve prolapse), aortic valve disease -- 3. Rhythm - irregular heart rhythms frequently seen in women -- 4.

Muscle - causes for heart muscle failure -- ch. 3. Phases of a woman's life: what are the heart risks? -- 1. Younger than forty-five: the "Honeymoon" period - 2. Ages forty-five to sixty-five: the beginning of change and menopause -- 3. Older than sixty-five: aging and the female heart -- ch.4. Are you woman at risk for heart disease? -- 1. Does age impact your risk? -- 2. Do race and ethnicity impact your risk? -- 3. How does family history affect your risk? -- 4. The danger of diabetes in women -- 5. Tobacco can make a woman's life go up in smoke -- 6. Hypertension: the silent killer of women -- 7. Are women treated appropriately for high cholesterol? -- 8. Obesity: the growing risk factor for women of all ages -- 9. Is your diet a heart health hazard? -- ch. 5. How do you know if you have a heart problem? -- 1. Are a woman's symptoms different from a man's? -- 2. Know the symptoms -- 3. What are the risk factors for women? -- 4. Take action against heart disease -- 5. Cardiac rehabilitation -- ch. 6. The heart of the matter on hormone therapy and oral contraception -- 1. Hormone therapy -- 2. Oral contraception -- ch. 7. Medications to treat heart disease -- 1. Beta blockers -- 2. Diuretics -- 3. ACE inhibitors -- 4. Antiotensin II receptor blockers -- 5. Calcium channel blockers -- 6. Statins -- 7. Fibrates -- 8. Other lipid-altering medications -- 9. Aspirin -- 10. Other blood thinning medications -- ch. 8. Herbal and natural supplements: how effective and safe are they? -- 1. Overview -- 2. Black cohosh -- 3. Coenzyme Q10 (COQ10) -- 4. Ephedrine and ephedra (Ma huang) -- 5. Fish oil -- 6. Flaxseed -- 7. Garlic -- 8. Ginkgo biloba -- 9. Plant stanols and sterols -- 10. Red yeast rice -- 10. Soy protein -- 12. St. John's wort -- ch. 9. Stress, depression, and anger: is your heart at emotional risk? -- 1. The effects of stress on a woman's heart and health -- 2. Depression - often overlooked but a serious risk to women -- 3. Methods and treatments to handle stress -- ch. 10. Cardiac testing -- 1. What is the best test for women and why? -- 2. Electrophysiologic testing -- 3. Stress testing -- 4. Angiography -- ch. 11. Procedures and surgery: why women do more poorly than men -- 1. Angioplasty -- 2. Coronary artery bypass surgery -- 3. Valve surgery -- 4. Pacemakers and defibrillators -- 5. Heart transplantation -- 6. The artificial heart -- ch. 12. Genetics and the female heart: what you should know about your genetic legacy -- 1. Case vignette: three generations of aortic dissection -- 2. Case vignette: another family with aortic dissection -- 3. The importance of heredity -- ch. 13. Living with heart disease -- 1. Knowing your options -- 2. Accepting the diagnosis -- 3. Support from family members -- 4. Lifestyle modifications -- ch. 15. Nine more things you need to know -- 1. How are women protected from arteriosclerosis until menopause? -- 2. How about the "pill"? Is it really as bad as they say? -- 3. I heard on TV that my migraines might be caused by a hole in my heart. How can this be? -- 4. Does a woman need any special wound care after open heart surgery? -- 5. I don't want the incision down the middle of my chest. What alternatives do I have? -- 6. What are the dangers of pregnancy on the heart? -- 7. What is peri-partum cardiomyopathy? -- 8. What is peri-partum aortic dissection? -- 9. How does heart disease affect sexuality in woman? -- ch. 15. Prospects for the future -- Appendix: Increase your strength and stay healthy -- Notes -- Glossary -- Index. Subjects: Heart diseases in women--Popular works. Notes: Includes bibliographical references (p. 265-274) and index.

INDEX

C

D

E

F

G

H

I

J

K

L

M

N

O

S